W9-BZW-293

William C. Kyle, Jr. Joseph H. Rubinstein Carolyn J. Vega

A Division of The McGraw-Hill Companies
Columbus, Ohio

Authors

William C. Kyle, Jr.
E. Desmond Lee Family
Professor of Science Education
University of Missouri – St. Louis
St. Louis, Missouri

Joseph H. Rubinstein
Professor of Education
Coker College
Hartsville, South Carolina

Carolyn J. Vega
Classroom Teacher
Nye Elementary
San Diego Unified School District
San Diego, California

PHOTO CREDITS
Cover Photo: ©VIREO/Academy of Natural Sciences Philadelphia

SRA/McGraw-Hill
A Division of The **McGraw-Hill** *Companies*

Send all inquiries to:
SRA/McGraw-Hill
8787 Orion Place
Columbus, OH 43240-4027

Printed in the United States of America.

ISBN 0-02-683803-6

5 6 7 8 9 RRW 05 04 03

Content Consultants

Gordon J. Aubrecht II
Professor of Physics
The Ohio State University at Marion
Marion, Ohio

William I. Ausich
Professor of Geological Sciences
The Ohio State University
Columbus, Ohio

Linda A. Berne, Ed.D., CHES
Professor/Health Promotion
The University of North Carolina
Charlotte, North Carolina

Robert Burnham
Science Writer
Hales Corners, Wisconsin

Dr. Thomas A. Davies
Texas A&M University
College Station, Texas

Nerma Coats Henderson
Science Teacher
Pickerington Local School District
Pickerington, Ohio

Dr. Tom Murphree
Naval Postgraduate School
Monterey, California

Harold Pratt
President, Educational Consultants, Inc.
Littleton, Colorado

Mary Jane Roscoe
Teacher/Gifted And Talented Program
Columbus, Ohio

Mark A. Seals
Assistant Professor
Alma College
Alma, Michigan

Sidney E. White
Professor Emeritus of Geology
The Ohio State University
Columbus, Ohio

Ranae M. Wooley
Molecular Biologist
Riverside, California

Reviewers

Stacey M. Benson
Teacher
Clarksville Montgomery County Schools
Clarksville, Tennessee

Mary Coppage
Teacher
Garden Grove Elementary
Winter Haven, Florida

Linda Cramer
Teacher
Huber Ridge Elementary
Westerville, Ohio

John Dodson
Teacher
West Clayton Elementary School
Clayton, North Carolina

Cathy A. Flannery
Science Department Chairperson/Biology Instructor
LaSalle-Peru Township High School
LaSalle, Illinois

Cynthia Gardner
Exceptional Children's Teacher
Balls Creek Elementary
Conover, North Carolina

Laurie Gipson
Teacher
West Clayton Elementary School
Clayton, North Carolina

Judythe M. Hazel
Principal and Science Specialist
Evans Elementary
Tempe, Arizona

Melissa E. Hogan
Teacher
Milwaukee Spanish Immersion School
Milwaukee, Wisconsin

David Kotkosky
Teacher
Fries Avenue School
Los Angeles, California

Sheryl Kurtin
Curriculum Coordinator, K-5
Sarasota County School Board
Sarasota, Florida

Michelle Maresh
Teacher
Yucca Valley Elementary School
Yucca Valley, California

Sherry V. Reynolds, Ed.D.
Teacher
Stillwater Public School System
Stillwater, Oklahoma

Carol J. Skousen
Teacher
Twin Peaks Elementary
Salt Lake City, Utah

M. Kate Thiry
Teacher
Wright Elementary
Dublin, Ohio

Life Science

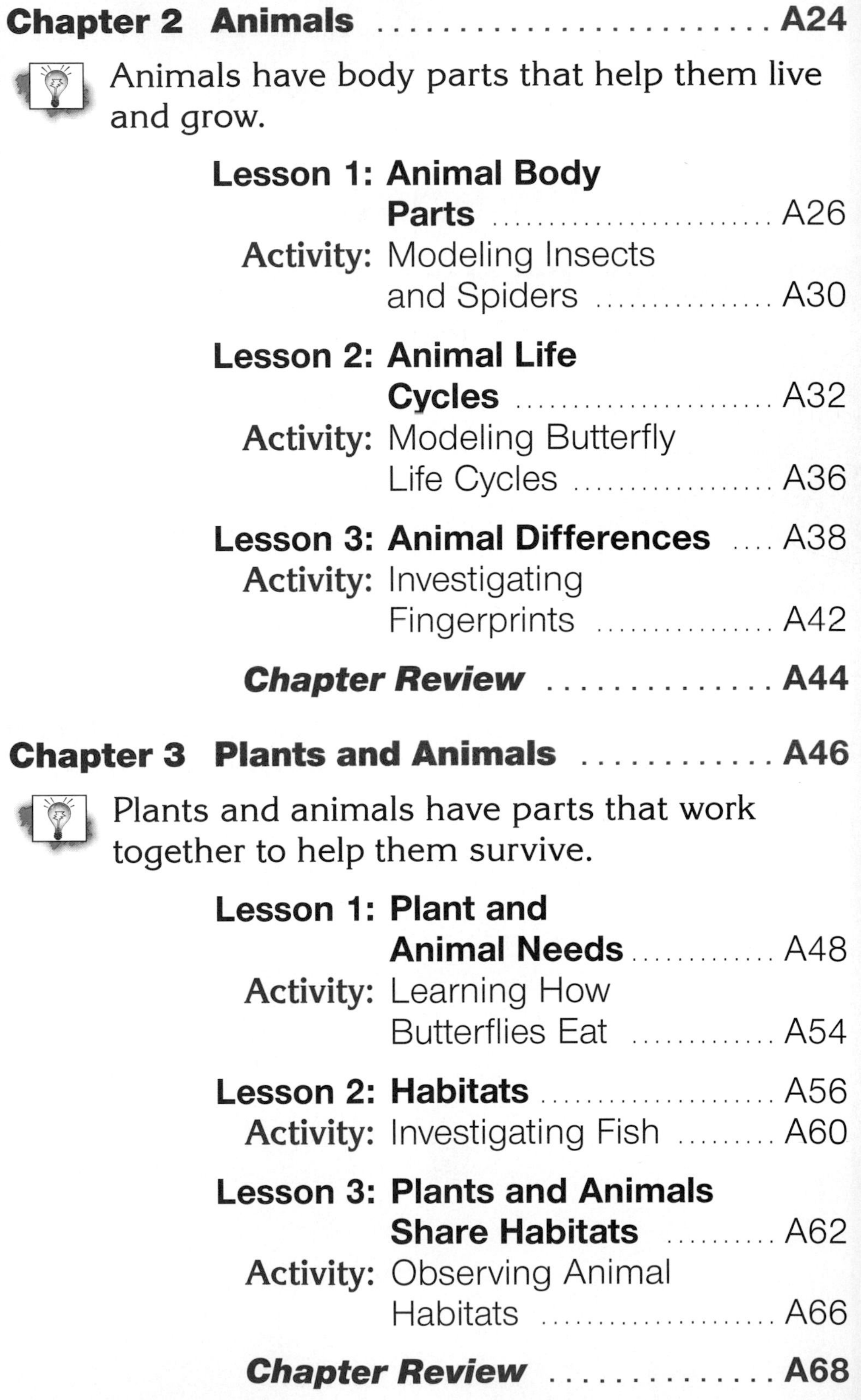

UNIT

B

Earth Science

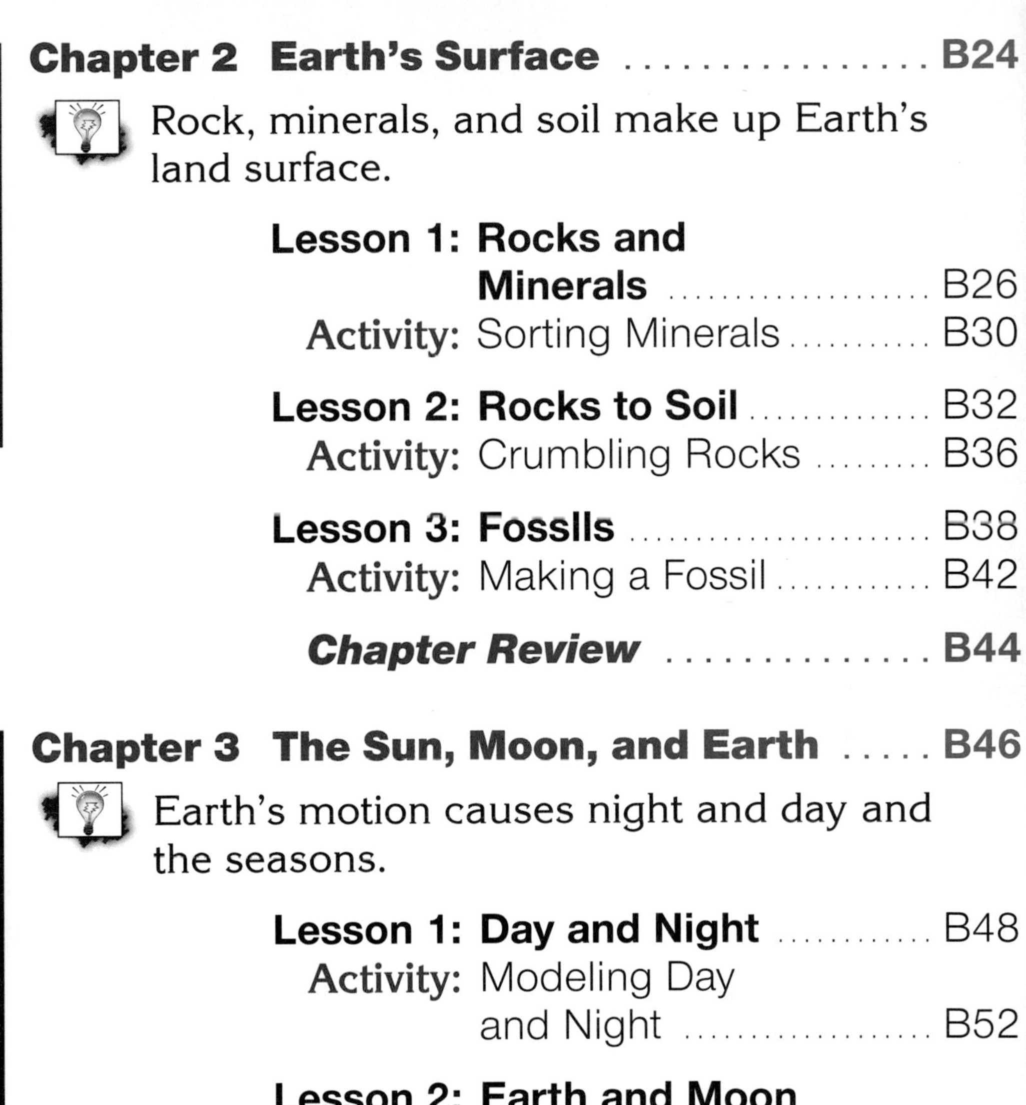

UNIT C

Physical Science

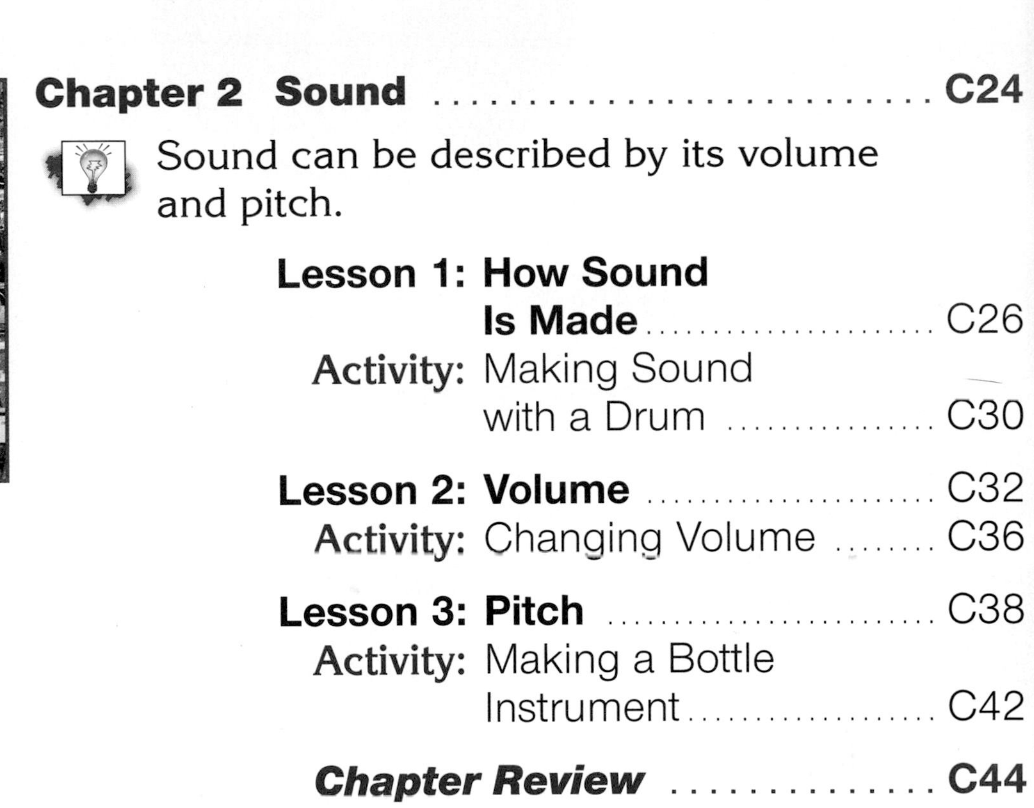

Health Science

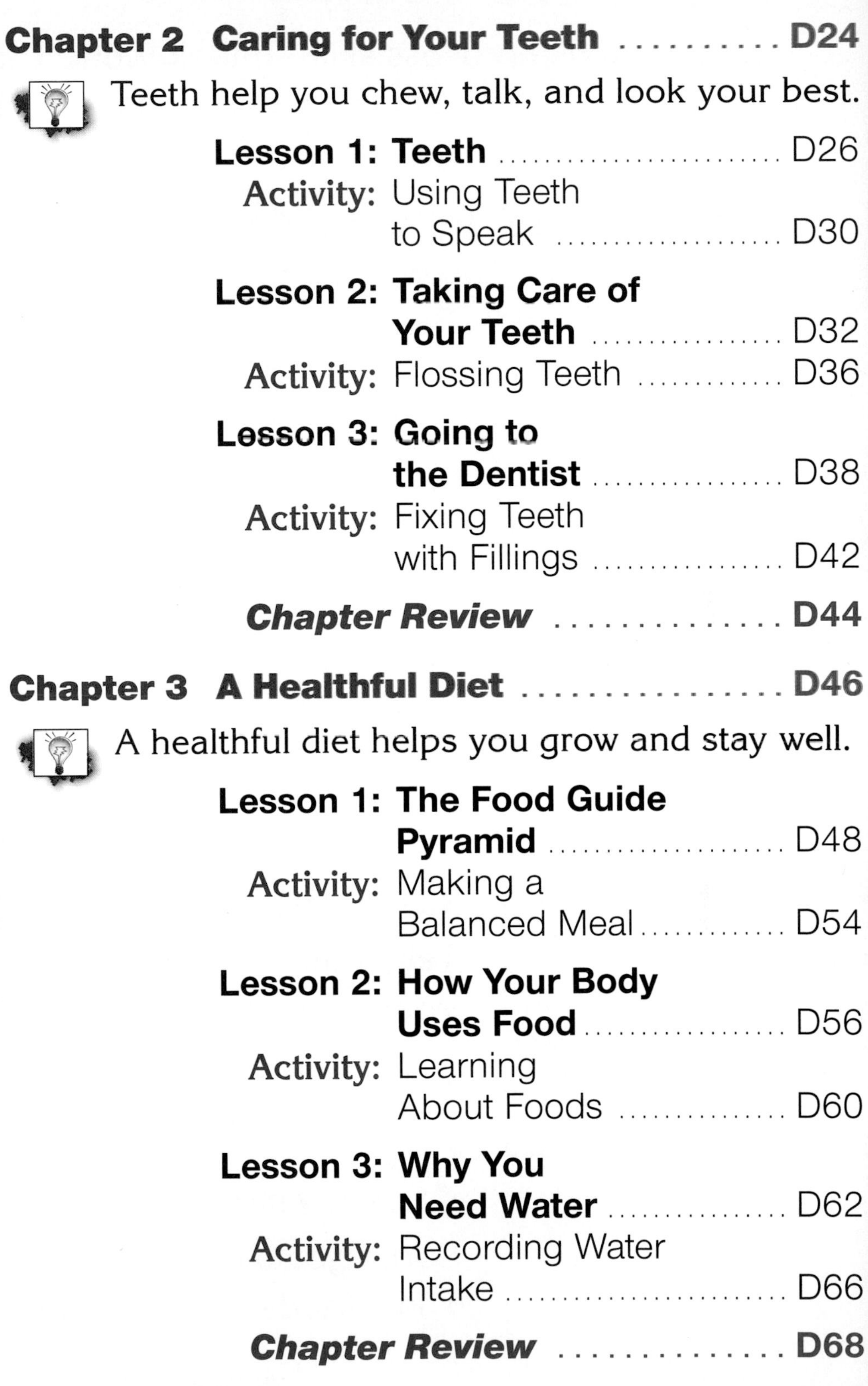

Science Process Skills

Scientists use process skills in their work. These skills help them to study science. Process skills also help them to discover new things.

Process skills will help you to discover more about science. You will use these skills in your science activities. Read about each skill. Think about how you already use some of these skills every day. Did you have any idea that you were such a scientist?

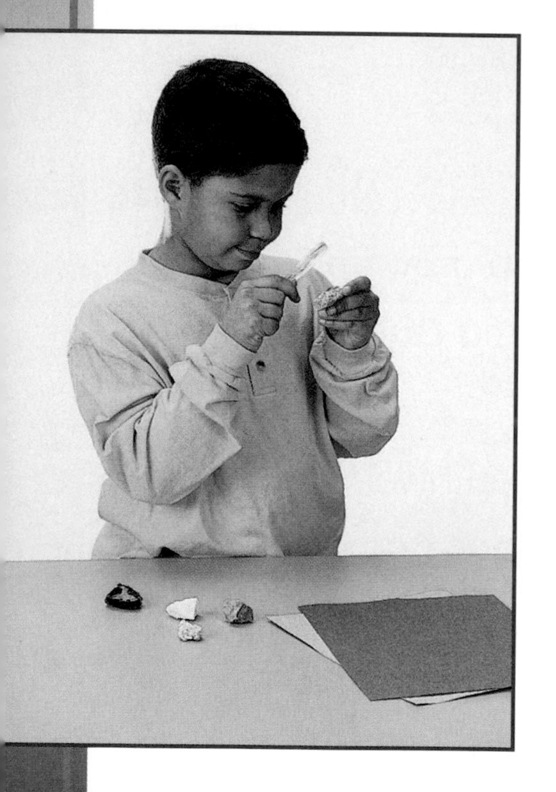

Observing

Find out about objects and events using your senses. You observe by seeing, hearing, touching, tasting, and smelling.

Looking at a rock with a hand lens is observing.

Communicating

Tell others what you know by speaking, writing, drawing, or using body language.

Drawing a picture of a frog is a way of communicating in science.

Classifying

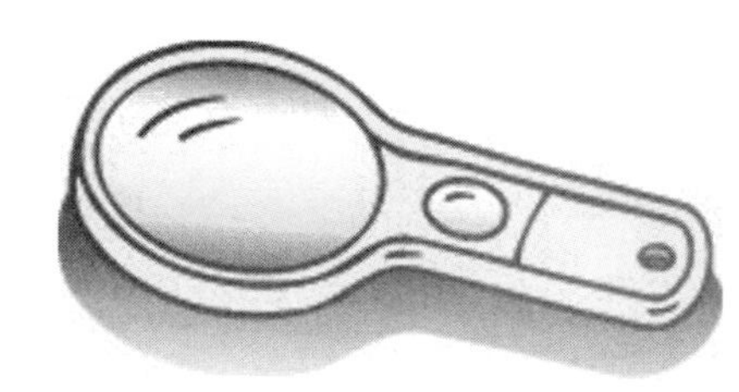

Sort objects and events into groups. The sorted objects should all be alike in some way.

Sorting crayons by their colors is a way of classifying.

Using Numbers

Use math skills to help understand and study science.

Counting the number of rainy days in a month is using numbers.

Measuring

Use measures of time, distance, length, size, weight, volume, mass, and temperature to compare objects or events. Measuring also includes using standard measurement tools to find answers.

Measuring skills include **using a stopwatch** to find out how many seconds it takes two worms to travel a distance of 20 cm.

Constructing models

Draw pictures or build models to help tell about thoughts or ideas or to show how things happen.

Constructing models of spiders and insects helps you learn about them.

Inferring

Use what you observe to help explain why something happened or will happen.

You could **infer** that it is summer because the swimming pool is crowded with children.

Predicting

Tell what you expect to happen in the future. Predictions are based on earlier observations.

You know that juice bars can melt. You could **predict** that a juice bar will melt if left in the sun.

Interpreting Data

See patterns or explain the meaning of information that you collect.

You are interpreting data when you **record** the temperature for one week and **answer** questions about whether it is hot or cold outside.

Identifying and Controlling Variables

Change one thing to see how it affects what happens.

You can **control** the amount of light plant leaves get. Covering some of the leaves on a plant with foil allows you to observe how plant leaves react to light.

Hypothesizing

Tell how or why something happens. Test the hypothesis to see if it is true or false.

You can **say** plants need water to grow. You have to **test** it with an experiment before you can say it is true.

Defining Operationally

Tell what something is by describing what you observe or what something does.

Saying soil is something plants grow in is a way to define soil operationally.

Designing Investigations

Plan investigations to gather data that will support or not support a hypothesis. The design of the investigation determines which variable will be changed, how it will be changed, and what type of data is expected.

You can **design an investigation** to test how sunlight affects plants. Place one plant in the sunlight and an identical plant in a closet. This will allow you to control the variable of sunlight.

Experimenting

Do an investigation to get information about objects, events, and things around you.

Experimenting pulls together all of the other process skills.

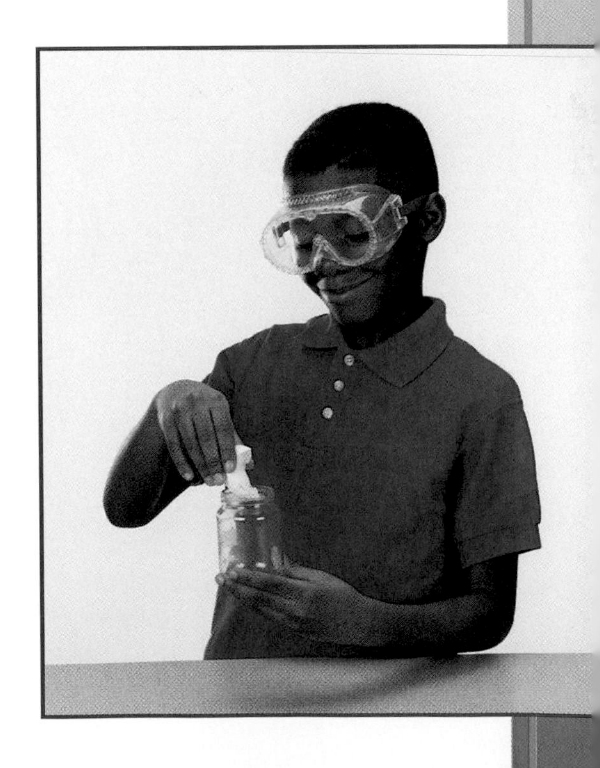

UNIT A

Life Science

Plants

There are many kinds of plants. Plants grow to many sizes, shapes, and colors. Pine trees and oak trees are plants. Tulips and daffodils are plants. What do you know about plants?

The Big IDEA

Plants grow and change.

CHAPTER
SCIENCE
INVESTIGATION
Set up an experiment
to see what seeds
need to grow. Find out
how in your Activity
Journal.

Lesson 1

Plant Parts

Let's Find Out

- Which plant parts make and store food
- Which plant parts make new plants

Words to Know

leaves
roots
nutrients
stem
flowers
seeds

The Big QUESTION

How do different parts of plants help them meet their needs?

Making Food

Plants have different parts that work together to help them get what they need. Green **leaves** make food in the plant. Green plants use sunlight, air, and water to make food in their leaves.

Tell about the plants in the picture. What kinds of plants grow where you live?

Roots get water from the soil. Roots also get **nutrients** from the soil. Plants need nutrients to make food. Most roots grow underground. They hold the plant in place.

Water moves from the roots to the leaves through the plant's **stem.** The stem is like a straw. Water goes up the stem into the leaves. Food from the leaves goes down the stem to the roots.

A carrot stores food in its root. It uses the food to grow and make new plants.

Water and nutrients enter through the roots and travel through the stem to the leaves. Food travels from the leaves to the roots.

Making New Plants

Plants also have special parts that make new plants. In many plants, these parts are called **flowers.**

Flowers make fruit and **seeds.** Flowers can be many colors, shapes, and sizes. Flowers grow on many kinds of plants. Find the flowers in these pictures.

Inside the seed is a tiny new plant and food to help it grow. Young plants use this food when they begin to grow. Beans, sunflower seeds, and peas are all seeds. With water, soil, and sunlight, they can grow into new plants.

CHECKPOINT

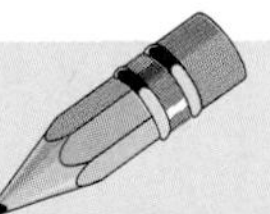

1. What plant parts help the plant make food?
2. What plant parts make new plants?

How do different parts of plants help them meet their needs?

ACTIVITY

Investigating Leaves

Find Out
Do this activity to see how water moves through a leaf.

Process Skills
- **Observing**
- **Communicating**
- **Inferring**
- **Predicting**

WHAT YOU NEED

large tree leaf

food coloring

cup of water

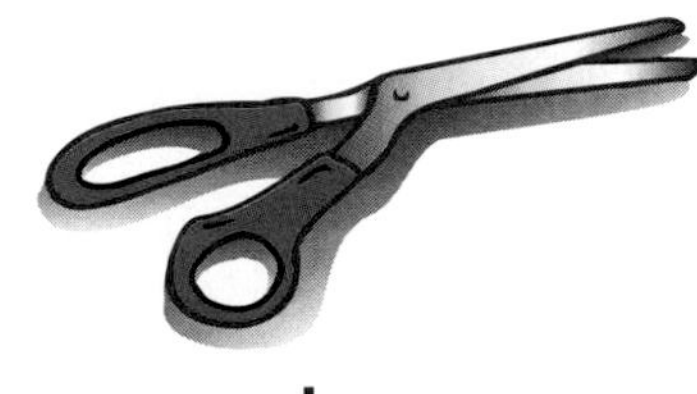

scissors

Activity Journal

WHAT TO DO

1. Watch your teacher put food coloring in the water.
2. Cut off the end of the leaf's stem.
   Safety! **Be careful with scissors.**

3. Place the leaf in the cup of water. Leave it overnight.
4. **Observe** the leaf.
5. **Draw** what happened.

What Happened

1. How did your leaf change?
2. Why do you think that your leaf changed?

What If

Predict what would happen if you did this activity with a flower.

Plant Life Cycles

Let's Find Out

- How flowering plants make seeds
- How plants without flowers grow new plants

Words to Know

life cycle
pollen
embryo
fruit
germinate
spores

The Big QUESTION

How do plants grow?

Flowers and Seeds

All living things go through a **life cycle.** The life cycle is how a living thing begins its life, grows, and makes new living things like itself.

Plants that grow flowers make **pollen.** Sometimes pollen needs to go to a new flower before it can make a seed.

After pollen gets into the flower, the **embryo** begins to grow. An embryo is a new, young plant. It is very small.

A seed grows. The embryo is part of the seed. The embryo is protected and has food to help it grow.

And something grows around the seed—a **fruit.** The fruit protects the seed.

Pollen helps flowers grow seeds.

When the seed gets what it needs, it will germinate.

Tiny roots grow down.

Not all seeds will begin to grow. Seeds need water. Some seeds need cold weather. Some seeds need warm weather. When the seed gets what it needs, it will **germinate.** The embryo in the seed begins to grow.

When it is an adult, the plant will make flowers. The flowers will make seeds. Then the seeds will make new plants. The new plants will continue the life cycle.

A tiny shoot grows up.

The new plant grows bigger.

Plants Without Flowers

Some plants like pine trees do not make flowers. Their seeds are in cones, not in fruit.

Pine trees have leaves shaped like needles. The leaves stay green all year long.

Some plants do not make seeds. They do not have flowers or cones. Ferns and mosses make **spores.** Spores fall on the ground. New plants grow in the soil.

Ferns

A moss is a small green plant with no flowers that grows close to the ground. Mosses grow in damp, shady places.

A fern has no flowers and has spores on the bottom side of its leaves. Most ferns grow where it is moist and shady.

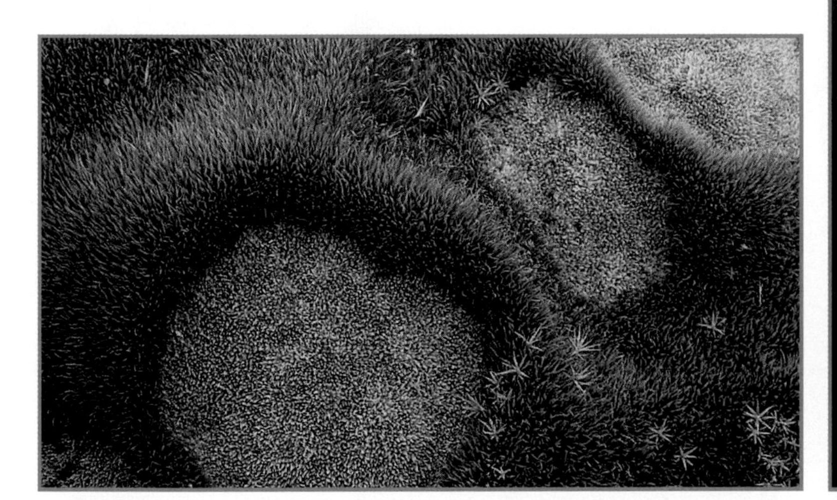

Mosses

CHECKPOINT

1. How do flowering plants make seeds?
2. How do plants without flowers grow new plants?

 How do plants grow?

ACTIVITY

Observing Bean Seeds

Find Out

Do this activity to observe and describe the parts of a bean seed.

Process Skills

Observing
Communicating
Predicting
Inferring

WHAT YOU NEED

bean seeds

cup of water

hand lens

paper towel

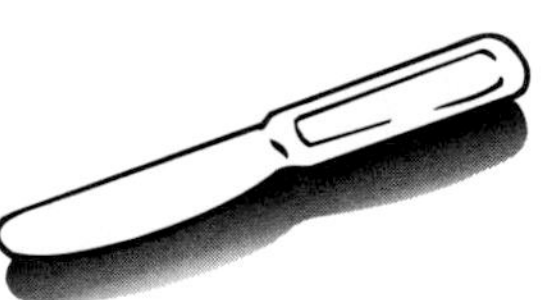

plastic knife

Activity Journal

WHAT TO DO

1. Soak the bean seeds in a cup of water overnight.

2. Peel off the skin of one seed. Pull the second seed apart by splitting it down the middle, lengthwise.
3. **Observe** the inside of the opened seed with the hand lens. Find the embryo. **Draw** what you see.
4. **Predict** what sunflower seeds would look like on the inside.
5. From what you have learned about seeds, **draw** pictures that show how a seed germinates.

What Happened

1. What did the inside of the seed look like?
2. What do you think was in the seed around the embryo?

What If

Predict what would happen if you planted half of a seed.

Plant Differences

Let's Find Out

- How the same kinds of plants can be different
- How crowding changes the way plants grow

Words to Know

variations
crowding

Variation in Plants

It is easy to see that maple trees look different from cactus plants. But if you look closely, maple trees are different from each other. Maple trees have leaves that are shaped the same. But some maple trees have more leaves than others.

Daisies usually have the same kind of flowers. But daisy plants may have a different number of flowers. Differences between the same kinds of living things are called **variations.**

The roots of a carrot plant are not the same as a cactus root. But all carrots have the same type of roots. All carrots have the same type of leaves. Carrots have variations too. Some carrots are large. Some carrots are small.

Not all seeds from a plant will grow exactly the same. Some seeds never germinate. Some seeds germinate quickly. Some seeds take a little longer to germinate.

Crowding Plants

To grow, plants need enough light and water. If many plants are close together, not all of them will grow. This is called **crowding.** If there is not enough light and water, some seedlings may be small and weak.

Maple trees growing with many other trees will be tall and straight with few branches. Maple trees growing in open spaces will have wider branches, and may not grow very tall. How the maple tree grows depends on the space around it and the amount of water and sunlight it receives.

Without enough water, most plants do not survive.

Farmers plant corn in rows. This gives the corn space to grow so that the corn will grow tall. If the seeds are planted too close together, they do not get enough water and sunlight. The corn crop will be small.

CHECKPOINT

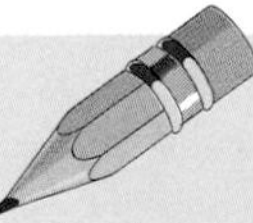

1. In what ways are the same kind of plants alike? In what ways can they be different?
2. How can crowding change the way plants grow?

Why don't all plants grow the same way?

ACTIVITY

Observing Leaves

Find Out

Do this activity to observe and describe variation among leaves.

Process Skills

Observing
Communicating
Measuring
Classifying
Interpreting Data

WHAT YOU NEED

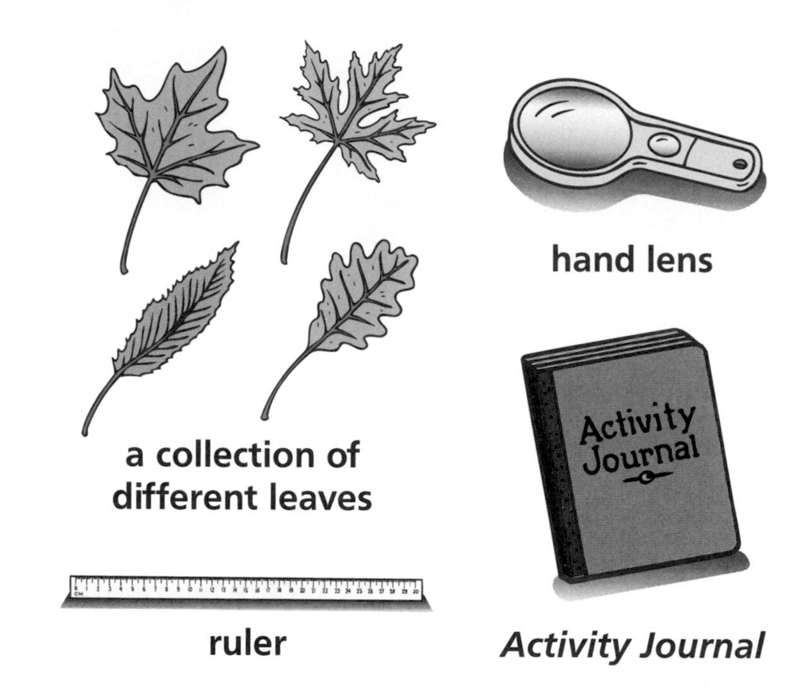

WHAT TO DO

1. Collect some leaves from different types of plants.
2. Collect some leaves from the same type of plant.

Safety! **Watch out for plants that might hurt you, like poison ivy or nettles. Be careful not to take too many leaves from one kind of plant.**

3. Use the hand lens to **observe** the leaves. **Draw** what you see.
4. **Measure** the leaves.
5. Compare the leaves. **Group** the leaves in different ways.
6. **Make a graph** to show the sizes of the leaves you collected.

What Happened

1. What were the differences in size of the leaves you collected?
2. How many different shapes of leaves did you find?

What If

What other plant parts could you measure and compare?

Review

What I Know

Choose the best word for each sentence.

germinate	**leaves**	**crowding**
fruit	**pollen**	**roots**
flowers	**seeds**	**variations**
nutrients	**spores**	**stem**
embryo	**life cycle**	

1. The __________ of a plant grow down into the soil.

2. Water goes up the __________ and into the leaves.

3. Pollen, seeds, and fruit are made in the __________.

4. The __________ have an embryo in them and food to help them grow.

5. The __________ protects seeds.

6. Ferns and mosses make __________.

7. Differences between the same kinds of living things are called __________.

Using What I Know

1. What plant parts can you see in the picture?

2. What plant parts cannot be seen?

3. How are the plants alike? How are they different?

For My *Portfolio*

Think about the different kinds of plants that grow where you live. Use pictures from old magazines to make a collage of plants that you have seen.

There are many kinds of animals. Fish, birds, and reptiles are animals. Amphibians, mammals, and insects are animals. Animals are alike in many ways. Each kind of animal is different, too.

What kind of animal do you see in this picture?

The Big IDEA

Animals have body parts that help them live and grow.

CHAPTER SCIENCE INVESTIGATION
Mealworms change as they grow. Take care of mealworms and watch the changes. Find out how in your ***Activity Journal.***

LESSON 1

Animal Body Parts

Let's Find Out

- What body parts insects have
- What kinds of animals do not have backbones

Words to Know

insects
backbone
thorax
abdomen
antennae

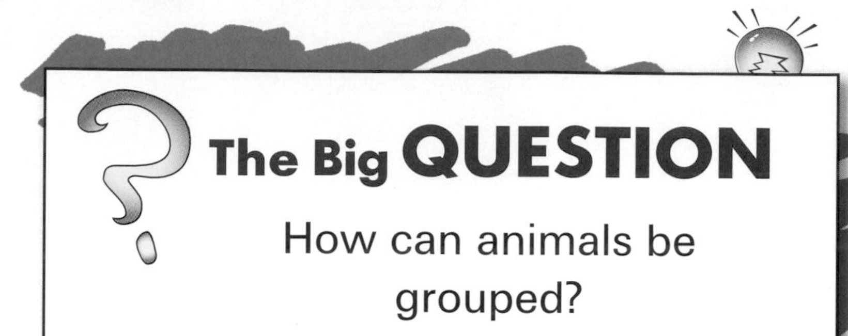

Insect Bodies

Animals have body parts that help them meet their needs. How many legs does this animal have? How does it use its legs?

Water strider

Insects are one kind of animal. Adult insects have some body parts in common. Insects have a thin shell on the outside of their bodies. This protects the parts inside them because insects do not have a **backbone**.

Insects have bodies with three main parts. They have a head, a **thorax** in the middle, and an **abdomen** at the end.

Insects use the **antennae** (an ten′ ē) attached to their heads to learn about the world around them. To help them move, insects have six legs that bend. Many insects use wings to fly.

These two insects have the same body parts, but the parts look different and move in different ways. The fly moves its small wings very fast. The butterfly moves its large wings more slowly. Look at the body parts in these pictures.

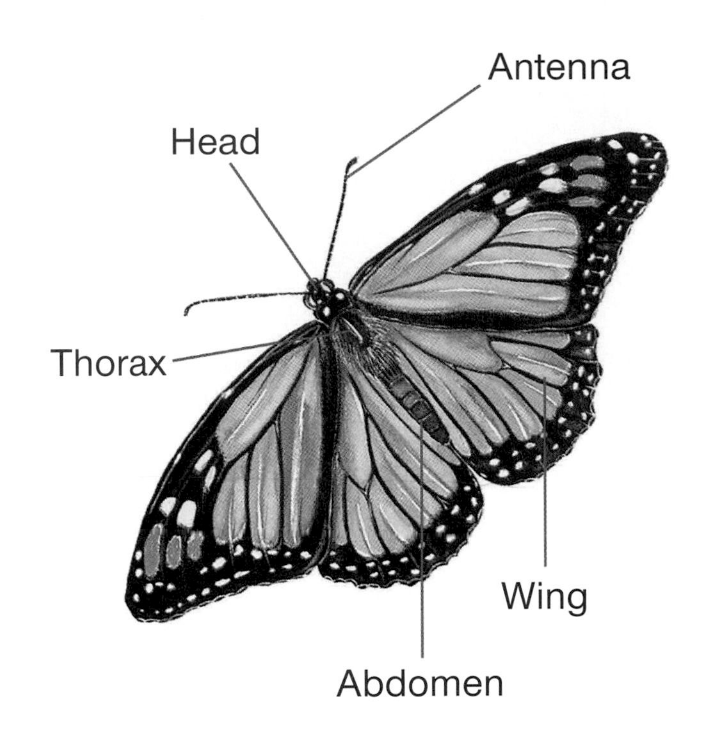

Butterfly

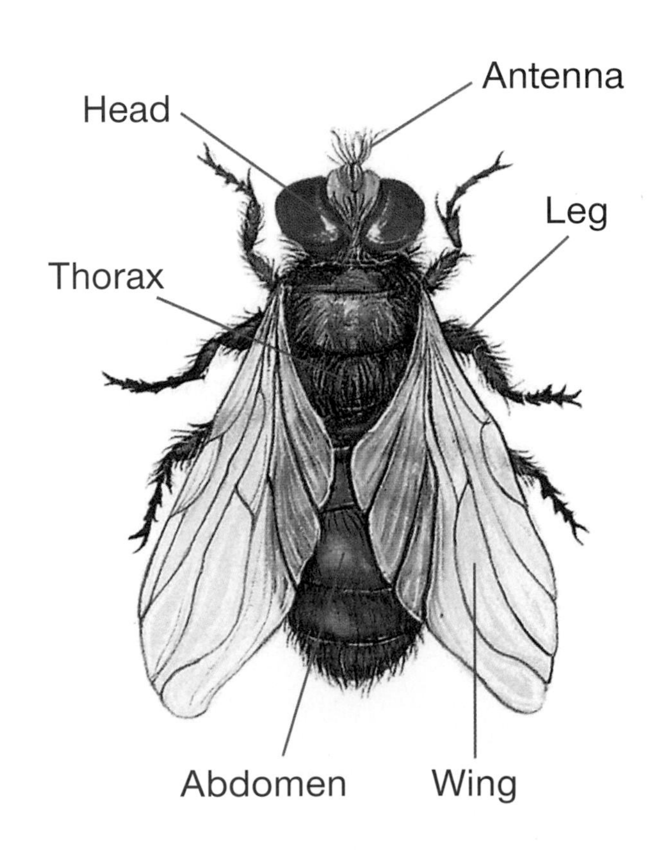

Fly

Other Animals Without Backbones

Spider

Crab

Centipede

Millipede

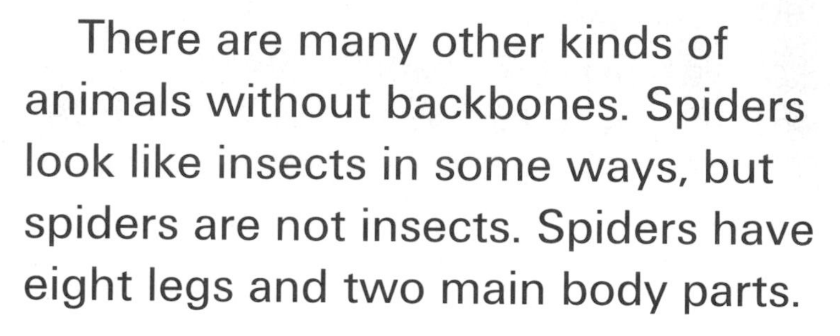

There are many other kinds of animals without backbones. Spiders look like insects in some ways, but spiders are not insects. Spiders have eight legs and two main body parts.

Crabs have a shell like insects. Crabs have ten legs. Two of the legs end in claws.

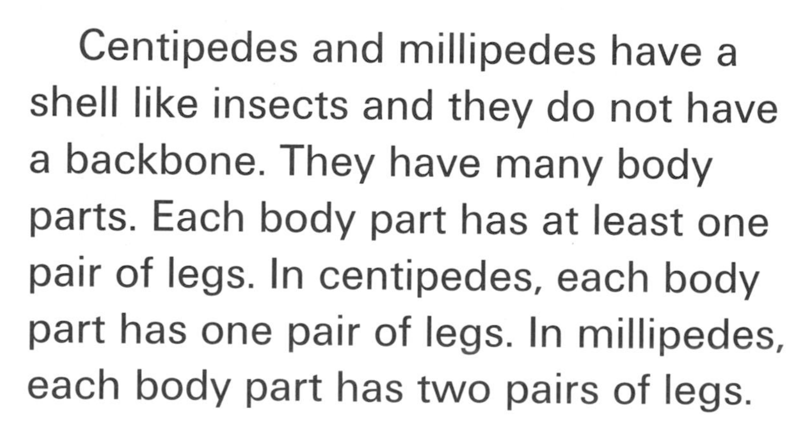

Centipedes and millipedes have a shell like insects and they do not have a backbone. They have many body parts. Each body part has at least one pair of legs. In centipedes, each body part has one pair of legs. In millipedes, each body part has two pairs of legs.

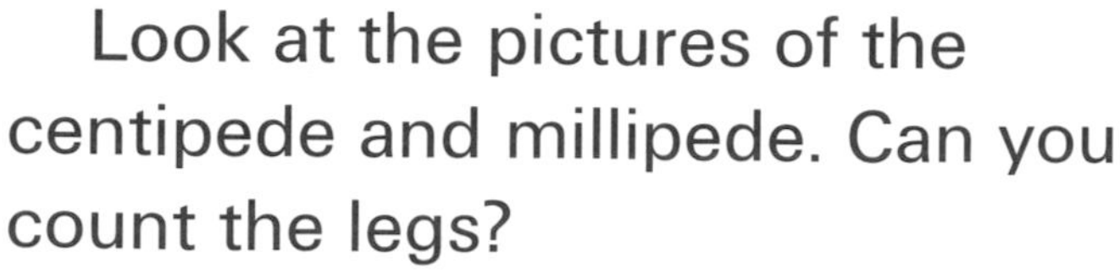

Look at the pictures of the centipede and millipede. Can you count the legs?

Starfish

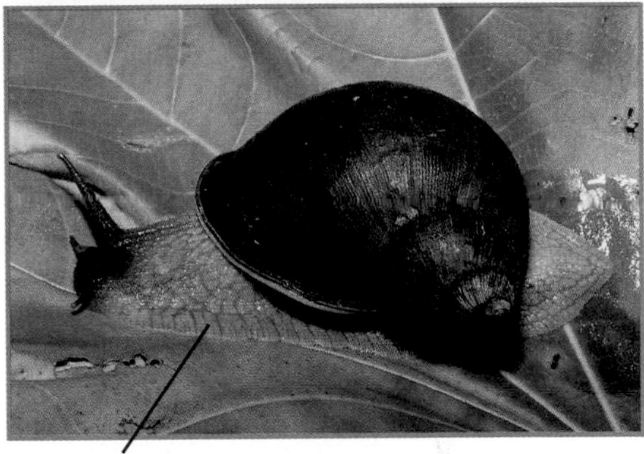

Foot Snail

Some animals without backbones do not use legs to move. Snails move by sliding over slime made by a large foot. Snails have a hard shell that protects their soft bodies.

Starfish live in tide pools. Starfish have hundreds of tiny tube feet. These feet work like tiny suction cups. Starfish use them to crawl and to hold onto rocks.

CHECKPOINT

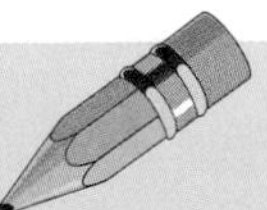

1. What body parts do insects have?
2. What are some animals that do not have backbones?

 How can animals be grouped?

ACTIVITY

Modeling Insects and Spiders

Find Out

Do this activity to observe the differences between insects and spiders.

Process Skills

Constructing Models
Observing

WHAT YOU NEED

modeling clay

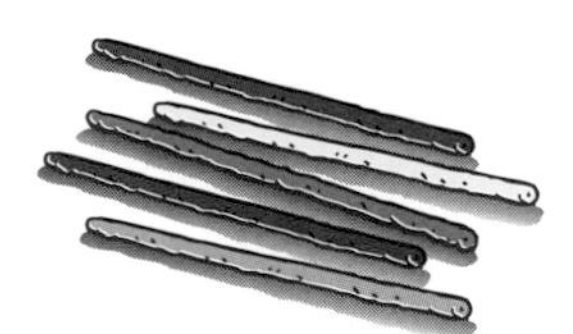

craft pipe cleaners

newspaper

Activity Journal

WHAT TO DO

1. List the parts of an insect's body. List the parts of a spider.
2. Cover the table with newspaper to set up a work space.

3. Use the clay and craft pipe cleaners to **model** an insect and a spider. Show all the parts.
4. **Observe** the two models.
5. Compare your models with those of a partner.

WHAT HAPPENED

1. How are your insect model and spider model alike?
2. How are they different?

WHAT IF

How would insects and spiders be different if they had only two legs?

Animal Life Cycles

Let's Find Out

- How different kinds of animal young are alike and different
- How insects change as they grow

Words to Know

larva
pupa

Animal Young

Different kinds of animals have different life cycles. Some animals take only a few weeks to grow. Other animals take many years to grow.

Humans are mammals. When human babies are born they look like adult humans. They have the same body parts. They have arms, legs, eyes, ears, and mouths. But these body parts are small.

Human babies cannot take care of themselves. Adult humans take care of them until they are grown. Humans take many years to grow.

Birds lay eggs with hard shells. When young birds hatch, they cannot take care of themselves. Many young birds hatch with their eyes closed. They cannot fly and do not have feathers. Parent birds must care for the young until they can care for themselves.

This robin feeds its young a caterpillar. After a few weeks, the young robins will grow feathers like their parent and will be able to fly. One day they will have young of their own.

Most fish, reptiles, and amphibians hatch from eggs too. After these snakes hatch, they will take care of themselves. They will find their own food and water. They will grow bigger and one day have young of their own.

Insect Life Cycles

Some animals do not become adults simply by growing larger. Most young insects look very different from adult insects. That is because insects go through different stages in their lives. They have different body parts in each stage.

The mealworm beetle in the picture hatched from an egg as a **larva.** The larva spends most of its time eating and sheds its skin as it grows.

The larva stops eating and becomes a **pupa.** The pupa does not move. It is changing into an adult beetle. The changes take many weeks. Finally an adult beetle crawls out of the pupa case.

Caterpillar

Chrysalis

A butterfly also goes through different stages in its life. It hatches from an egg as a caterpillar. After it grows bigger it sticks itself to a branch. Its skin splits open. Underneath is a case called a chrysalis. In time the case opens. The butterfly unfolds its wings.

Butterfly

CHECKPOINT 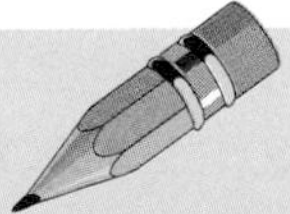

1. How are different kinds of animal young alike and different?
2. How do insects change as they grow?

 How do animals grow?

ACTIVITY

Modeling Butterfly Life Cycles

Find Out
Do this activity to learn the different stages of a butterfly's life cycle.

Process Skills
Constructing Models
Communicating

WHAT YOU NEED

newspapers

modeling clay

crayons

scissors

construction paper

Activity Journal

WHAT TO DO

1. Cover your work area with newspapers.
2. Use the clay to **make** a caterpillar.

3. Use the clay to **make** a chrysalis.
4. Fold a sheet of construction paper in half.
5. **Draw** an outline of a butterfly wing.
6. Cut out the butterfly wings and unfold the paper.

   Safety! Be careful with scissors.
7. **Draw** the body of the butterfly on the butterfly wings and color it with the crayons.

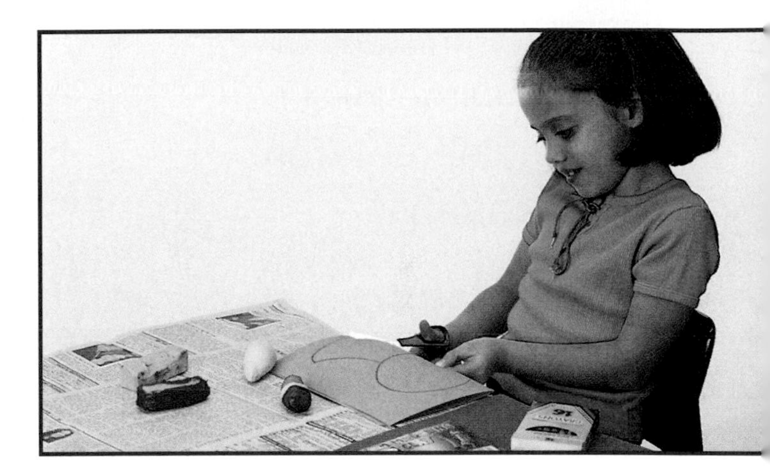

What Happened

1. How are the caterpillar and chrysalis alike? How are they different?
2. How are the caterpillar and butterfly alike? How are they different?

What If

Suppose you observed both a butterfly and a caterpillar. Compare how they move from place to place.

LESSON 3

Animal Differences

Let's Find Out

- How animals look like their parents
- How animals of the same kind can be different from each other

Words to Know

variations

The Big QUESTION

How are animals different?

Parents and Young

People are alike in many ways. People have arms, legs, heads, and hair. People have eyes, ears, mouths, and noses. But, every person looks different. Some people are short, and some people are tall. Some people have brown hair, and some have red hair.

These differences are called **variations.** Babies grow up to look like their parents. If the parents are both tall, the children might be tall. But no one knows exactly how tall they will be. In a family, the people might look alike in some ways, but each person looks different in some ways too.

How do the members of this family look alike and different?

When they are grown, most animals look like their parents. Dalmatian puppies look alike in some ways. They are all covered in fur. They are all white with black spots. They all have four long legs and a long, thin tail.

The puppies look different from each other in some ways too. The puppies might have more or fewer spots. The puppies might grow to be large or small.

How do these puppies look different?

Variations in Animals

From far away these pigeons look alike. They all have beaks, wings, and feathers. Up close you can see the differences. Some are darker than others. Some are smaller than others.

Dragonflies are insects. They each have six legs and a head, abdomen, and thorax. Dragonflies have two pairs of wings.

Each dragonfly looks different, though. Some dragonflies are bright and colorful. Some are blue and green. Some are brown.

Tiger moths are usually black, white, and red. But each tiger moth looks different. There are variations in the color and the pattern on the wings.

CHECKPOINT

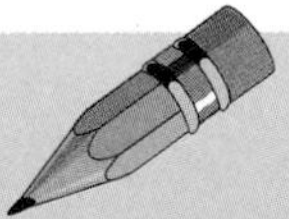

1. In what ways do animals look like their parents?
2. What kinds of variations do animals have?

How are animals different?

ACTIVITY

Investigating Fingerprints

Find Out
Do this activity to see how human fingerprints are alike and different.

Process Skills
Observing
Predicting

WHAT YOU NEED

paper **pencil** **clear tape**

hand lens

Activity Journal

WHAT TO DO

1. Rub the pencil on a piece of paper many times to darken a small area.
2. Rub the tip of one finger on the dark area. **Observe** the

tip of your finger with your hand lens. Cover the darkened fingertip with a piece of clear tape. **Predict** what your fingerprint will look like.

3. Press the tape against your finger and then carefully lift off the tape. Place the tape onto another piece of paper and write your name. Repeat the steps with another finger.
4. **Observe** your fingerprints with a hand lens.
5. Compare your fingerprints with your classmates' fingerprints.

What Happened

1. Did your fingerprints look the same?
2. How were your fingerprints different from your classmates' fingerprints?

What If

What would happen if you made your fingerprints on the same piece of tape as someone else?

Review

What I Know

Choose the best word for each sentence.

pupa	thorax	abdomen
insects	backbone	variations
larva	antennae	

1. __________ have six legs and three main body parts.

2. Insects, spiders, snails, and starfish do not have a __________.

3. Insects have a head, a __________, and an abdomen.

4. __________ help insects learn about the world around them.

5. A mealworm beetle hatches from an egg as a __________.

6. The __________ will change into an adult insect.

7. Differences like hair color and eye color are __________.

Using What I Know

1. Which kinds of animals do you see in these pictures?

2. How can you tell which animal is an insect?

3. In what ways are the animals alike? In what ways do they look different?

For My Portfolio

Write a story about a butterfly's life. Draw pictures to show how the butterfly looks during each stage of its life.

Plants and Animals

Plants and animals live together in many different kinds of places. They have parts that work together to help them live.

Some parts help plants make food. Some parts help animals find food. Other parts protect plants and animals. These parts keep them safe.

The Big IDEA

Plants and animals have parts that work together to help them survive.

CHAPTER SCIENCE INVESTIGATION
Watch ants live and work. Find out what to do in your *Activity Journal.*

Lesson 1

Plant and Animal Needs

Let's Find Out
- What parts help plants grow
- What body parts help animals survive
- What body parts help some animals live under water

Words to Know
- **survive**
- **proboscis**

The Big QUESTION
How do plants and animals get what they need?

Helping Plants Grow

Plants and animals have parts that help them live and grow. They help living things **survive.**

Pufferfish

Elephant

Plants need nutrients to live and grow. Plant roots pull needed nutrients from the soil. Roots also hold plants in place. Roots help plants survive.

In the rain forest, soil is not deep. Some rain forest trees have big roots above ground that keep the tree from falling.

Plants need to spread their seeds far apart. Dandelion seeds are very light. Just a puff of wind can blow them far. Maple seeds are shaped like helicopter wings. The wind blows the seeds into the air. The shape of the seeds helps them move.

Animal Survival

Animals have body parts to help them survive. A giraffe's long neck allows the giraffe to eat the leaves from the branches of tall trees. This gives the giraffe food that other animals cannot reach.

Proboscis

Butterflies have a **proboscis** (prə bäs′ əs) that allows them to get food from deep inside flowers. The proboscis is a hollow tube, like a straw.

Most turtles have hard shells. The shells protect their bodies. Some turtles can even pull their heads inside of the shell to try to stay safe.

Can you find the turtle's head?

Living in Water

Some animals have body parts that help them live in water. Fish live in the water. Fish use fins to move through the water. Fish have gills. Gills take the air that fish need out of the water.

Lobsters live on the bottom of the ocean. Lobsters have gills too. Lobsters have a tail shaped like a fan. This shape allows the lobster to swim.

This octopus has eight arms with suckers. The arms help the octopus to move. The suckers help octopuses get food. They allow the octopuses to grab fish and crabs.

CHECKPOINT

1. What parts help plants grow?
2. What body parts help animals survive?
3. What body parts help animals live under water?

? How do plants and animals get what they need?

ACTIVITY

Learning How Butterflies Eat

Find Out

Do this activity to find out how a butterfly drinks from flowers.

Process Skills

Constructing Models
Inferring
Communicating

WHAT YOU NEED

two straws

paper

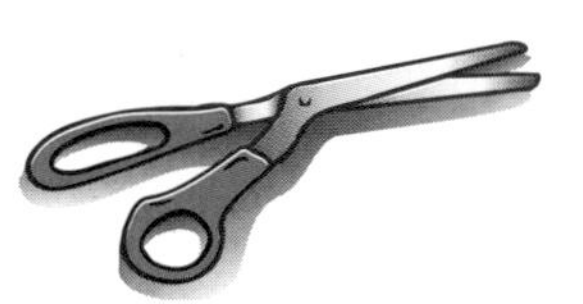

scissors

cup of water

crayons

hole punch

Activity Journal

WHAT TO DO

1. Fold the paper in half. Draw the outline of a butterfly wing and head.
2. Cut out the butterfly.
   Safety! **Be careful with scissors.**

3. Use crayons to color your butterfly's wings, body, and head.
4. Use a hole punch to make a hole in the butterfly's head.
5. Put two straws together to make one long straw. This is the proboscis.
6. Put the proboscis through the hole in your butterfly.
7. Bring your butterfly to the water. Pretend you are the butterfly. Drink from the straw.
8. Curl up the straw when you are finished. This is how a butterfly curls up its proboscis.

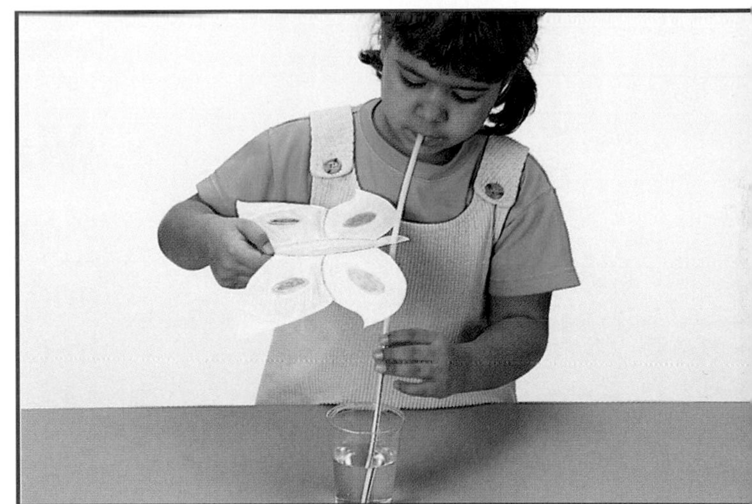

What Happened

1. What part of the butterfly did you use to drink the water?
2. **Tell** how this body part helped you drink.

What If

How do you think the proboscis would look if butterflies did not get food from deep inside flowers?

Habitats

Let's Find Out

- What kinds of plants and animals live in different habitats
- How color, size, and shape protect plants and animals

Words to Know

habitats

The Big QUESTION

How do plants and animals survive in their habitats?

Many Habitats

The different places where plants and animals live are called **habitats.** In cold habitats, some plants are small. Mountain wildflowers grow close to the ground where it is warmer and less windy.

Bears that live in cold places eat a lot of food before winter comes. In winter, bears cannot find food to eat. They sleep much of the time to save energy.

Many bears spend winter in a den lined with dry leaves.

A pond is another habitat. Pond plants need parts to get sunlight and air. Water lilies have big leaves that float on the top of the water and take in air and sunlight to make food.

Water lilies

Animals that live in a pond need parts to swim with. The water boatman is an insect that lives on the water. It has legs that are shaped like oars in a boat. These legs push the insect across the water.

Water boatman

Blending In

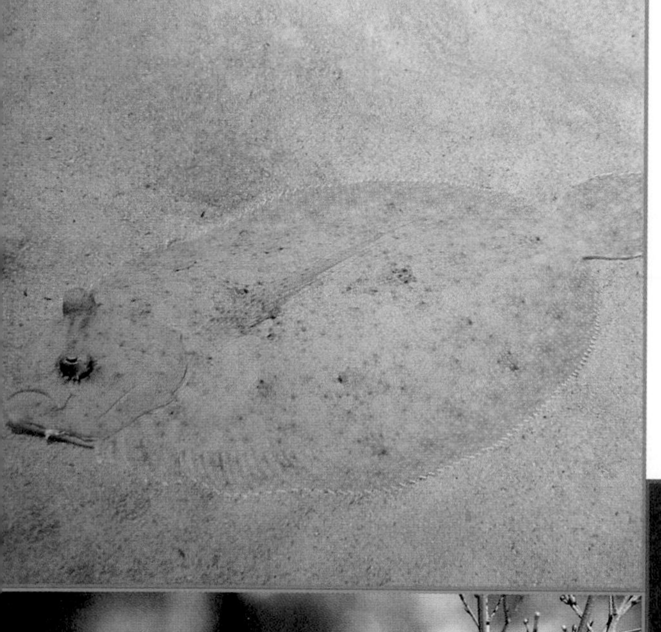

Flounder

Flounder live on the sandy ocean floor. They are the same color as sand. Their color allows the flounder to hide from other animals in search of food.

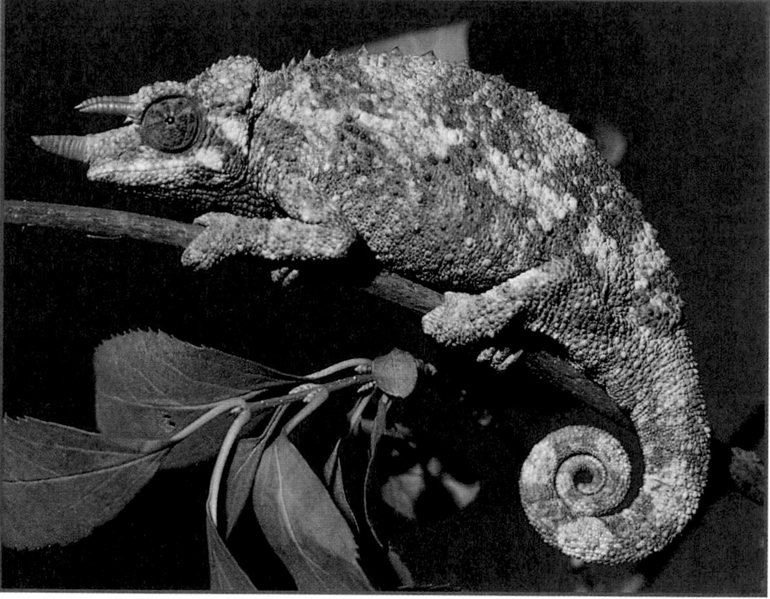

Some animals change color and can blend in with different places. Chameleons become darker when a larger animal comes near them.

It may take many minutes for chameleons to change color from green to dark.

Many insects are colored so that they blend in with their habitat. Their coloring and shape make it hard for other animals to find and eat them. It also keeps them hidden from other insects that they try to catch. Can you find the katydid in this picture? Its coloring and shape make it hard to see.

A few plants have shapes and colors so that they blend into their habitat. Living stones are tiny plants with round leaves. They look like stones on the ground. Their shape makes it hard for animals to find and eat them.

CHECKPOINT

1. How are plants and animals that live in cold habitats different from those that live in a pond?
2. How do some animals hide from other animals?

How do plants and animals survive in their habitats?

ACTIVITY

Investigating Fish

Find Out

Do this activity to see how a fish's shape helps it move through water.

Process Skills

Constructing Models
Observing
Communicating
Inferring

WHAT YOU NEED

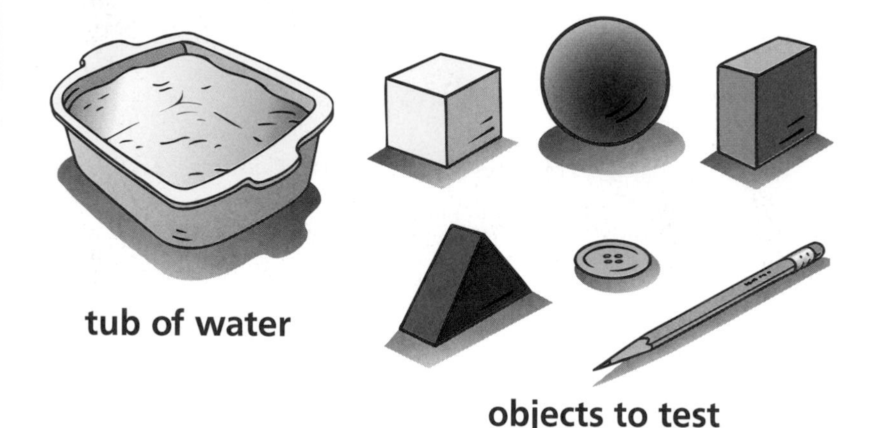

tub of water

objects to test

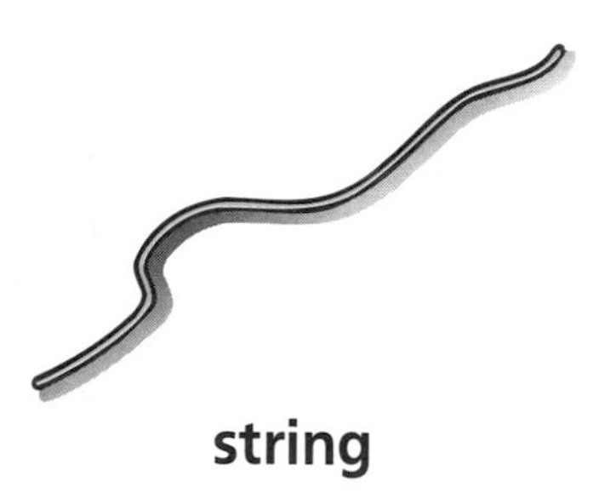

string

Activity Journal

WHAT TO DO

1. Tie a piece of string to one object.
2. Place the object near one end of the tub. Pull the object through the water. **Watch** what happens.

3. Tie strings to other objects and pull them, too. **Watch** what happens.
4. **Record** what you see.

What Happened

1. Which objects were easy to pull through the water?
2. Which objects were not easy to pull?

What If

What would happen if you pulled a star-shaped block through the water? Would it be easier or harder to pull than the square block of wood?

Plants and Animals Share Habitats

Let's Find Out
- How plants and animals protect themselves
- How plants and animals depend on each other

Words to Know
defenses
poison
mimicry

The Big QUESTION

How do plants and animals share habitats?

Animal Defenses

All living things need food to survive. Many animals eat plants. Many animals eat other animals. Some plants and animals have **defenses** that keep other animals away.

Poison ivy has **poison** on its leaves, stems, and roots. Animals do not eat it because it will make them sick. Even touching it can make you itch.

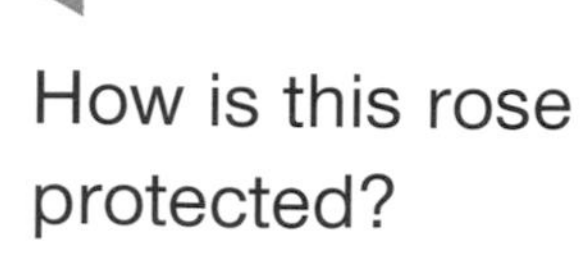

How is this rose protected?

Poison ivy

This horned lizard uses a special trick to stay alive. It puffs up with air when a large animal is close by. This might make it look too big for the animal to eat. If this doesn't work, it can squirt blood from slits above its eyes. This might startle the other animal. Then the horned lizard can escape by burrowing into the sand until only its eyes show.

Horned lizard

Some animals are protected because they look like more dangerous ones. This defense is called **mimicry.**

These two butterflies look alike. The monarch butterfly tastes bad to birds. But the viceroy butterfly does not. Birds learn that that black and orange butterflies taste bad. They do not eat viceroy butterflies.

Monarch butterfly

What body parts protect this porcupine?

Viceroy butterfly

Depending on Each Other

Some animals live and work in groups. Ants work in large groups. They build tunnels and live together. When an ant finds food, it carries the food home and shares it with other ants.

Honeybees live together in hives made of wax. Some honeybees collect food and bring it back to the hive. Others care for the young inside the hive.

Prairie dogs live together in prairie dog towns. One prairie dog stands guard. If there is danger it makes a noise that warns the others.

Plants and animals depend on each other. Many insects and birds drink nectar from flowers. When they do so, the pollen from flowers sticks to them. When they go to another flower, the pollen falls off. Then, seeds can be made and new plants can grow.

These horses have burrs stuck in their manes.

Some seeds stick to animals' fur. The seeds fall off in a new place. Because of this, plants grow in new places.

Some animals live in or near plants. Many baboons climb trees at night. They sleep in the trees. It is safer there than on the ground.

CHECKPOINT

1. How do plants and animals protect themselves?
2. How do animals and plants depend on each other?

? How do plants and animals share habitats?

ACTIVITY

Observing Animal Habitats

Find Out
Do this activity to find out where animals live.

Process Skills

Observing
Communicating

WHAT YOU NEED

crayons

paper

Activity Journal

WHAT TO DO

1. Go outside. Walk around your school.
2. **Look** for places where animals live. Some will be easy to see. Others may be hard to see. **Look** in trees and other plants and in places on the ground.

3. Every time you see a place where an animal lives, **draw** a picture of the habitat.
4. Name the animals that you see.

   Safety! **Avoid animals and plants that might be harmful. Try not to disturb the area you are observing.**

What Happened

1. What kinds of animal habitats did you see?
2. What kinds of animals might live in the places you saw?

What If

How would the area you observed look different in winter?

Review

What I Know

Choose the best word for each sentence.

habitats	**mimicry**
defenses	**survive**
proboscis	**poison**

1. Plants and animals have parts that help them __________.

2. A butterfly uses its __________ to get food from flowers.

3. The different places where plants and animals live are called __________.

4. Some plants and animals have __________ that warn animals to stay away.

5. If animals eat plants that have __________ on the leaves, it can make them sick.

6. __________ is when an animal is protected because it looks like a more dangerous animal.

Using What I Know

1. How many different kinds of plants do you see in the picture?
2. How do the plants look different?
3. How many different kinds of animals do you see?
4. What body parts do the animals in the picture have to help them survive?

For My *Portfolio*

Draw a new kind of animal that could live in the ocean. Tell how it protects itself from other animals.

Unit Review

Telling About What I Learned

1. Plants grow and change. Tell how flowers help plants grow and change.
2. Animals have body parts that help them live and grow. Name three animals and tell how they grow and change.
3. Plants and animals have parts that work together to help them survive. Name two plants and two animals. What parts allow them to meet their needs?

Problem Solving

Use the picture to help answer the questions.

1. What plant and animal parts can you find?
2. How do the plant and animal depend on each other?

Something to Do

What plant or animal body part would you like to have? Imagine how your life would change if you had that part. Write a short story or draw a picture of what you would do.

UNIT B

Earth Science

Planet Earth

Imagine you are going on a trip to the moon. What should you bring with you? You would need food for your trip because plants and animals don't grow on the moon. You would need water because there are no lakes or rivers on the moon. You would even need to bring air to breathe. Living things need water, air, and food. Earth provides the resources living things need.

The Big IDEA

People use Earth's resources in many ways.

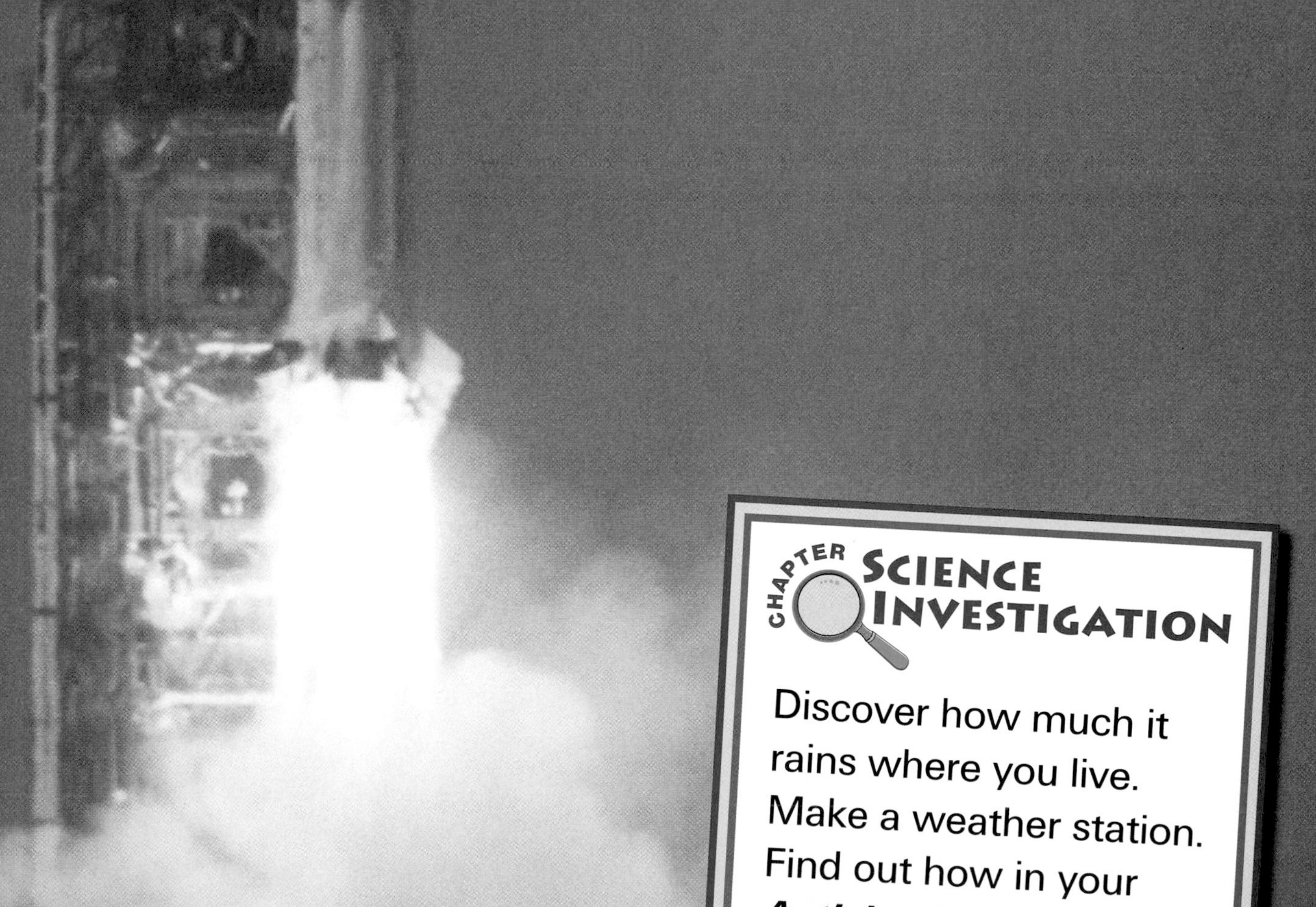

CHAPTER SCIENCE INVESTIGATION

Discover how much it rains where you live. Make a weather station. Find out how in your ***Activity Journal.***

LESSON 1

Earth

Let's Find Out

- Why Earth is sometimes called the blue planet
- How the water cycle works

Words to Know

natural resources
water cycle
water vapor

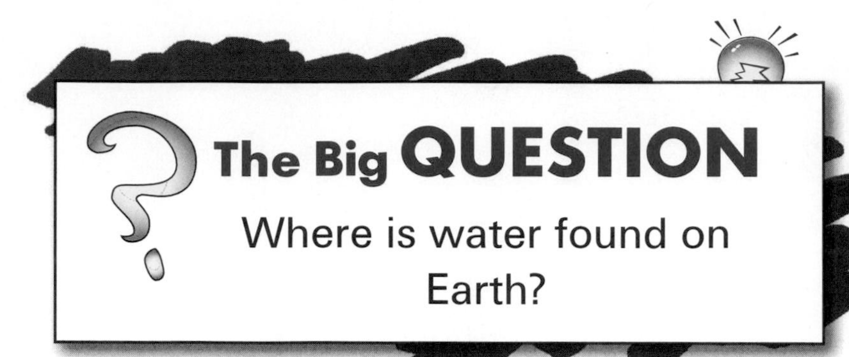

The Blue Planet

Natural resources are materials from Earth that plants and animals depend on for life. Rocks, soil, air, and water are all important natural resources.

Water may be Earth's greatest resource. Earth is sometimes called the blue planet because most of Earth is covered with water. When you look at Earth from space, you see many blue parts. The blue parts are liquid water. Water on Earth fills the oceans, lakes, rivers, and streams.

The oceans are full of salt water. Many plants and animals live in the oceans.

Streams, rivers, and most lakes are full of freshwater. Freshwater does not have much salt. Many kinds of fish live in the freshwater of streams, rivers, and lakes. Animals on land drink freshwater from streams, rivers, and lakes.

Most of Earth is covered by water.

The Water Cycle

Water is always moving in a cycle called the **water cycle.**

Clouds in the sky are made of liquid and frozen water. When it rains or snows, the water falls to Earth.

The sun shines on the water. Heat from the sun warms the water. Liquid water turns into a gas called **water vapor.**

Water vapor forms clouds. Water falls from the clouds as rain and snow. Rain and snow are made from freshwater. Rain and snow are never salty, even when the water comes from the ocean. The same water keeps rising into the sky as water vapor and falling back to Earth as rain. That's why it's called the water cycle. It happens over and over again.

CHECKPOINT

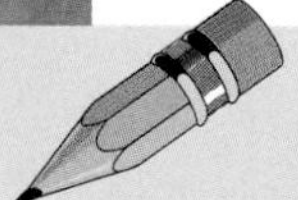

1. Why is Earth sometimes called the blue planet?
2. Tell how the water cycle works.

Where is water found on Earth?

ACTIVITY

Observing the Water Cycle

Find Out
Do this activity to see how the water cycle works.

Process Skills
Constructing Models
Observing
Communicating

WHAT YOU NEED

hot water

ice cube

small jar with lid

Activity Journal

WHAT TO DO

1. Watch your teacher pour the hot water into the jar. Close the lid.
   Safety! **Be careful with hot water.**

2. Wait a few minutes. **Observe** the sides of the jar and draw what you see.
3. Now, put an ice cube on top of the lid. **Observe** what happens and **record** what you see.

What Happened

1. What formed inside the jar?
2. Where did it come from?

What If

What would happen if you used salty water instead of tap water? Would the drops on the side be salty?

Earth's Natural Resources

Let's Find Out

- What kinds of things are made from natural resources
- What fossil fuels are and how they are used

Words to Know

energy
fossil fuels

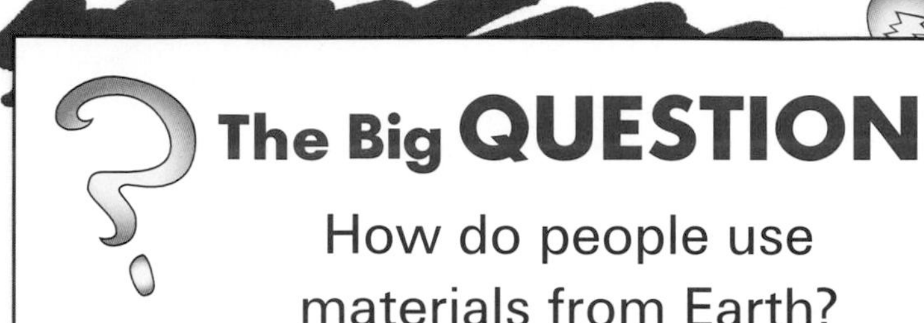

Made from Natural Resources

Imagine a bicycle made of rocks and trees, sunglasses made of oil, or windows made of sand. Many of the things you use every day are made from natural resources.

Bicycles are made of metal from rocks and rubber from trees.

Bicycles are made of metal and rubber. People get metal from rocks and some rubber from trees.

Sunglasses are made of plastic. People make plastic from oil.

Windows are made of glass. People make glass from sand.

The materials for things we use come from Earth. People change natural resources like rocks, soil, and water to make the things we need.

Sunglasses are made of plastic, which is made from oil.

Windows are made of glass, which is made from sand.

Fossil Fuels

Making things like bicycles, sunglasses, and windows takes energy. **Energy** makes things work. People can get energy from natural resources. **Fossil fuels** are natural resources. Some fossil fuels take millions of years to form.

People use fossil fuels like coal, oil, and natural gas for heat, electricity, and transportation. Factory machines use fossil fuels for power. Most cars use a fossil fuel too.

Some natural resources, like plants, grow faster than resources like coal or oil can form.

Factories, like this one that makes computers, depend on fossil fuels.

People use plants for food.

People use trees for building homes.

Plants and animals use natural resources too. Plants need soil, water, and air to grow. Animals need clean water to drink. Earth's materials are important for all living things.

People use trees to stay warm.

CHECKPOINT

1. What kinds of things are made from natural resources?
2. What are fossil fuels? How are they used?

How do people use materials from Earth?

ACTIVITY

Describing Earth's Resources

Find Out
Do this activity to see how Earth's resources can be classified.

Process Skills
Classifying
Communicating

WHAT YOU NEED

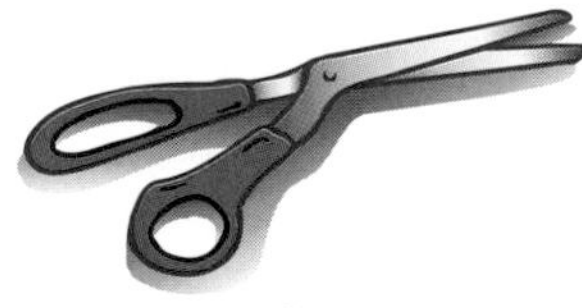
scissors

glue

magazines

Activity Journal

posterboard

WHAT TO DO

1. Cut out pictures of Earth's natural resources from magazines.

   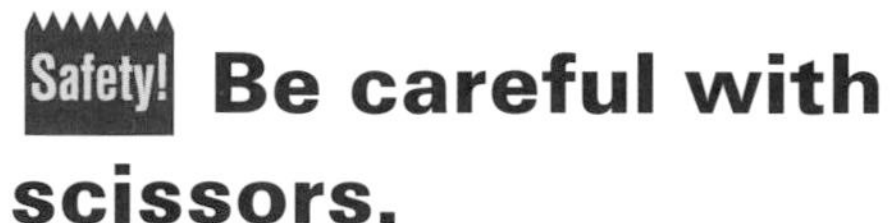
   Safety! **Be careful with scissors.**

2. Divide your posterboard into three parts. **Classify** your pictures into three groups:

A. Resources in nature (like water, soil, rocks, air, and trees)

B. Resources being changed (like in a factory or at a construction site)

C. Resources that have been changed into new materials (like plastic containers, houses, appliances, and food)

3. Glue the pictures from the three groups above into a collage.

4. Show your collage to a partner. **Tell** how each resource is changed to make something people need.

What Happened

1. What natural resources did you find?
2. **Tell** where you see fossil fuels being used in your collage.

What If

Draw links on your collage between resources and the materials made from the resources.

Conservation

Let's Find Out

- How people change the environment
- How people can take care of Earth

Words to Know

environment
pollution
conservation
reduce
reuse
recycle

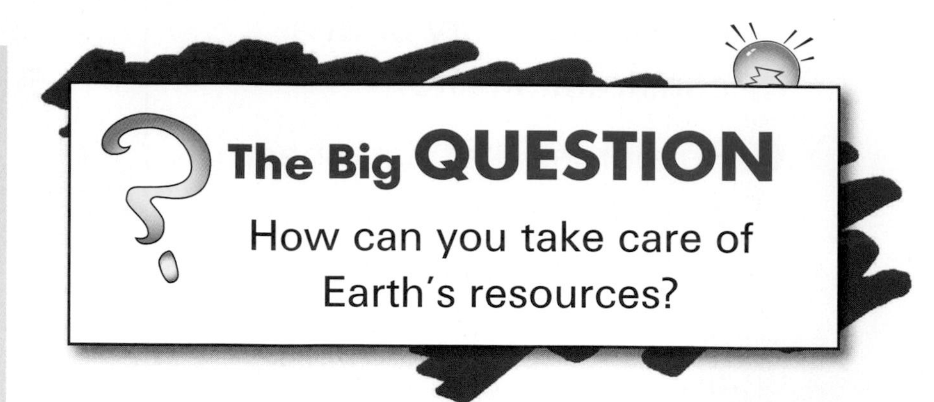

People Make Changes

Your **environment** is everything around you. Air, water, land, animals, and plants are all part of the environment. People change the environment in many ways.

People change the environment when they use natural resources. People dig mines to get coal out of the ground. They drill wells to get oil and natural gas. Sometimes people make

In strip mining, the land is dug away so that people can get coal and minerals. Companies try to make the land look like it did before it was mined.

many changes in the environment to get the resources they need.

Some changes people make to the environment cannot be undone. Air, water, and land can become polluted. Smoke and gases from cars and chimneys cause air pollution. **Pollution** makes our environment dirty. Pollution can be harmful to both plants and animals.

Plants and animals live in places where they can meet their needs. These places are called their habitats. Living things are adapted to their habitats. This means that when the habitats change, plants and animals cannot always adapt to the changes.

People can change habitats to meet their needs, so people can live almost anywhere. People build towns and farms where animals and plants live. Other animals and plants cannot easily adapt to changes. If animals cannot find another habitat nearby, they do not survive.

Taking Care of Earth

People use natural resources every day. People use natural resources to build and heat homes and to cook food to eat. But people can do these things in a way that takes care of Earth. People can practice **conservation**. Conservation means using resources wisely.

You can protect, or save, resources. Three ways to save resources are to reduce, reuse, and recycle.

Reduce means to use less. Turning off the lights when you leave a room reduces how much energy you use. Turning off the water while you brush your teeth reduces the amount of water you use.

Think twice before you throw something away. Can you use it for something else? When you use something again you **reuse** it. Reusing things reduces the amount of trash and the amount of resources people use.

One way to reuse things is to **recycle** them. If something is made of paper, aluminum, glass, or plastic, you can recycle it. Instead of throwing it in the trash, put it in a recycling bin. Recycled plastics can be used to make park benches, picnic tables, and toys. Recycling helps save resources that take a long time to replace. It also saves habitats for other living things.

There are many ways people can practice conservation. Think of ways you can help.

CHECKPOINT

1. How do people change the environment?
2. How can people take care of Earth?

How can you take care of Earth's resources?

ACTIVITY

Reusing Objects

Find Out

Do this activity to find out how to reuse objects.

Process Skills

Observing
Communicating

WHAT YOU NEED

crayons

scissors

objects to reuse

glue

Activity Journal

WHAT TO DO

1. **Look** at the objects. Think of ways that they could be used in a different way.

2. Choose one object. Create a new way to use it. You can use crayons, glue, and scissors to help you.

   Safety! **Be careful with scissors.**

3. **Explain** how you will use the object in a new way.

4. **Draw** a picture to show how you could use the object in another way.

What Happened

1. How did you use your object in a new way?
2. How could reusing your object help the environment?

What If

What other objects could be reused?

Review

What I Know

Choose the best word for each sentence.

natural resources	**water cycle**
water vapor	**recycle**
environment	**pollution**
conservation	**energy**
reduce	**fossil fuels**
reuse	

1. Water and soil are ________.

2. Water is always moving in the ________.

3. People use ________, like coal and natural gas, to heat their homes.

4. ________ makes land, water, and air dirty.

5. Using natural resources wisely is called ________.

6. Turning off the water when you brush your teeth will ________ the amount of water you use.

7. When you make something new out of something old, you ________ it.

Using What I Know

1. What kind of water do you see in the picture?
2. Where did the water come from?
3. Where would you find water vapor in the picture?

For My Portfolio

Write a story about a town that tries to conserve resources.

Earth's Surface

The surface of Earth is always changing. The ground you walk on is changing a little bit day by day. You can't see or feel Earth changing because it happens so slowly.

The Big IDEA

Rocks, minerals, and soil make up Earth's land surface.

CHAPTER SCIENCE INVESTIGATION
See how wind and water change Earth's surface. Find out how in your Activity Journal.

LESSON 1

Rocks and Minerals

Let's Find Out
- What makes up rocks
- How minerals can be described

Words to Know
rock
crust
minerals

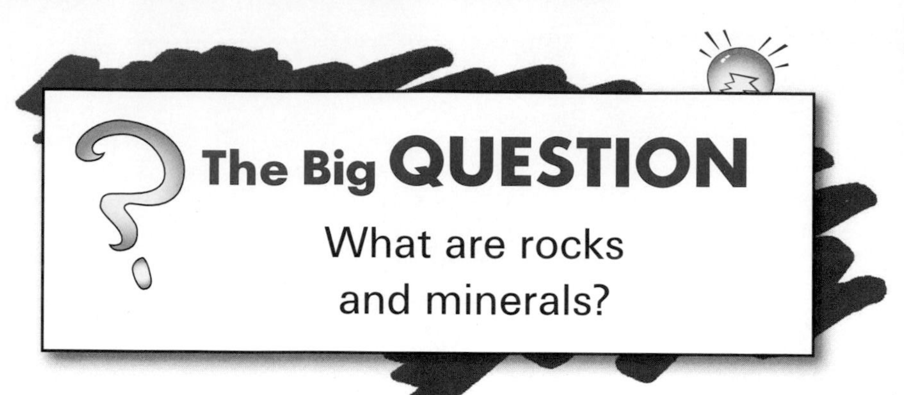

Rocks

A **rock** is a piece of Earth's **crust.** Earth's crust is the top part of Earth. Rocks can be as small as grains of sand or as large as boulders. Earth has many different kinds of rocks.

This geologist studies rocks.

Some rocks are sharp and pointy. Other rocks are round and smooth, like pebbles. Tell about the rocks in the picture.

Minerals

Pyrite

Chrysocolla

Malachite

Amethyst

Calcite

Copper

Rocks are made of **minerals.** Minerals are nonliving materials from Earth. Some rocks are made of only one mineral. But most common rocks are made of two or more minerals.

There are more than 500 kinds of minerals. Minerals can be dull or shiny, soft or hard. Minerals are many colors. Each mineral has a name. What a rock looks like depends on what minerals it is made of.

When you look at a rock you can name the minerals in it. Granite is a rock. Granite is made up of the minerals feldspar, mica, and quartz.

CHECKPOINT

1. What are rocks made of?
2. How can you describe minerals?

 What are rocks and minerals?

ACTIVITY

Sorting Minerals

Find Out
Do this activity to sort minerals into types.

Process Skills
Classifying
Communicating
Observing
Inferring

WHAT YOU NEED

hand lens

rock salt

quartz

nail

Activity Journal

goggles

WHAT TO DO

1. **Look** at the two minerals. **Record** your observations.
2. Take the nail and try to scratch the quartz. Look for a scratch mark.
   Safety! **Be careful with sharp objects.**

3. Try to scratch the salt with the nail. Look for a scratch mark in the salt.
4. **Observe** the minerals again with your hand lens.
5. **Draw** what you see.

What Happened

1. **Tell** which mineral was softer.
2. How do you know?

What If

Name another way you could classify the minerals.

Rocks to Soil

Let's Find Out
- How big rocks become sand
- What soil is made of

Words to Know
weathering
sand
soil
erosion

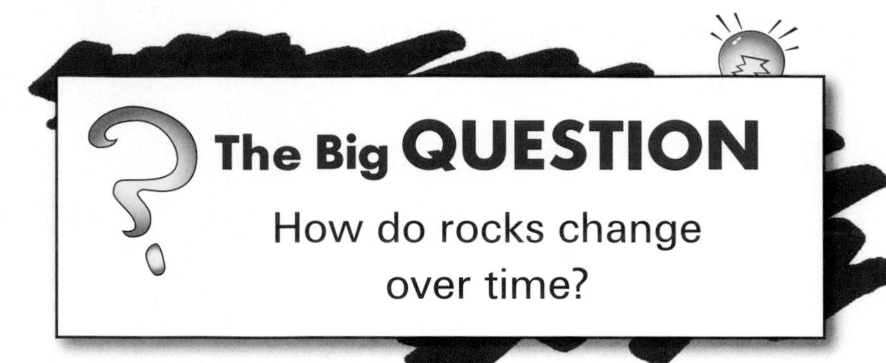

Rocks and Weathering

The surface of Earth is always changing. It changes very slowly. Wind, rain, cold, and heat change Earth's surface.

When wind and water break rock into smaller pieces, **weathering** occurs. What happens when waves crash onto a beach? Waves rub large and small rocks together. Over a long time, this makes the rocks smaller, rounder, and smoother. This is one type of weathering.

When wind blows sand against rock, the shape of the rock slowly changes. Wind is another type of weathering.

Everywhere on Earth, rocks are changing. Big rocks break into smaller rocks. Smaller rocks break into **sand.**

Soil

Soil takes a long time to form. Take a close look at soil and you will see why. You can see tiny rocks and sand mixed with dead plants and leaves. Soil is made of rocks, minerals, plants, and the remains of dead animals. Soil has water and air in it too.

Soil is not the same in every place. Soil can be different colors. Some kinds of soil are hard. Some kinds of soil are soft. Some kinds of soil are better for growing plants than others. New soil is always forming, just as Earth is always changing.

Rainwater, streams, and wind can move rocks and soil. A heavy rain can wash soil away. This is called **erosion.** Trees and plants protect soil. They keep it from blowing away. The roots of plants protect soil by slowing down erosion.

Erosion is a problem on farms, open fields, and hilly areas. Trees along the edges of farms and fields can help break the force of the wind. Then the wind doesn't blow as much of the soil away.

CHECKPOINT

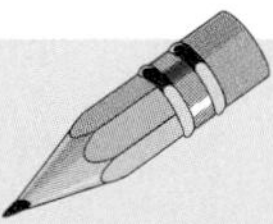

1. What are two types of weathering?
2. What is soil made of?

 How do rocks change over time?

ACTIVITY

Crumbling Rocks

Find Out
Do this activity to see the effects of weathering on rocks.

Process Skills
- Communicating
- Observing
- Predicting
- Inferring

WHAT YOU NEED

sheet of black construction paper

dry, empty milk carton

Activity Journal

sugar cubes

WHAT TO DO

1. Pretend the sugar cubes are rocks. Place six sugar cubes in the milk carton. Put one cube on the black paper.
2. Close the top of the milk carton and shake it 25 times.

3. Empty the carton onto the black paper next to the whole sugar cube.
4. **Draw** what the sugar cubes looked like before and after you put them in the carton.
5. Put four of the six sugar cubes back in the carton.
6. Repeat Steps 2 and 3, but this time shake the carton 50 times.
7. **Predict** how the sugar cubes will change and **draw** what you **observe.**

What Happened

1. How did the sugar cubes change?
2. How is this like what happens to rocks when they rub together?

What If

What would happen if you did this activity with rocks?

Lesson 3

Fossils

Let's Find Out

- How Earth's changing climate affected animals in the past
- How fossils form

Words to Know

climate
dinosaurs
fossils

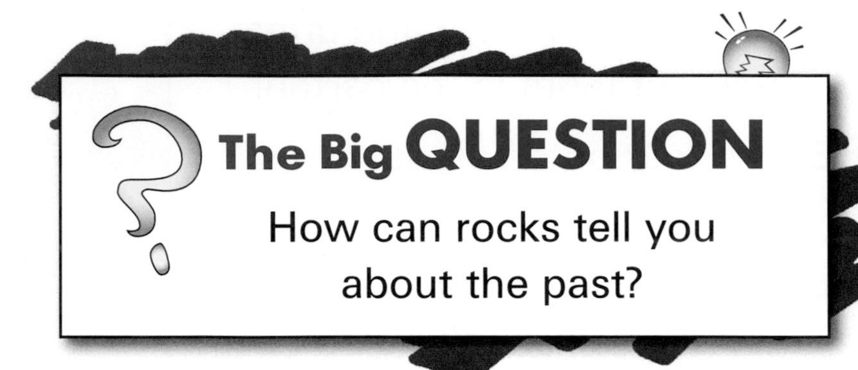

Changing Climate

Environments on Earth are not the same today as in the past. Earth's surface has changed over time. Earth's climate has changed too. **Climate** is what the weather of a place is usually like. Animals and plants are affected by changes in the climate.

Dinosaurs lived on Earth many millions of years before people.

Dinosaurs needed the same things that other animals need now. They needed food, water, and air.

Apatosaurus

Dinosaurs did not live alone on Earth. Insects, fish, reptiles, and some mammals shared Earth with dinosaurs.

No one knows for certain why dinosaurs died out. Many scientists think they died because a huge meteorite hit Earth. If it crashed into Earth, it probably caused a large cloud of dust. The dust blocked the sun's light. Earth became dark for a long time and the temperature changed. Many animals and plants could not adapt to the changes and died. Dinosaurs could not find the kind of food they needed to live.

Woolly mammoths

About two million years ago an ice age began on Earth. Parts of Earth were covered with ice. Animals like woolly mammoths lived in cold places.

Woolly mammoths also died out. Scientists do not know what happened to them. Some scientists think that when Earth's climate became warmer they could not adapt to the changes. Scientists study woolly mammoth skeletons that people have found frozen in the ice to try to find out what happened. What do you think happened to them?

Fossils

We know about dinosaurs because we have found their fossils. **Fossils** buried in rock layers give us clues about how animals and plants looked in the past. Fossils are evidence that an animal or plant once lived. They can be footprints, bones, seeds, or even dinosaur eggs.

Look at these pictures to see how some fossils are formed.

1. An animal or plant dies and is quickly buried or covered.

2. Over millions of years, the layers of rock build up, one upon another.

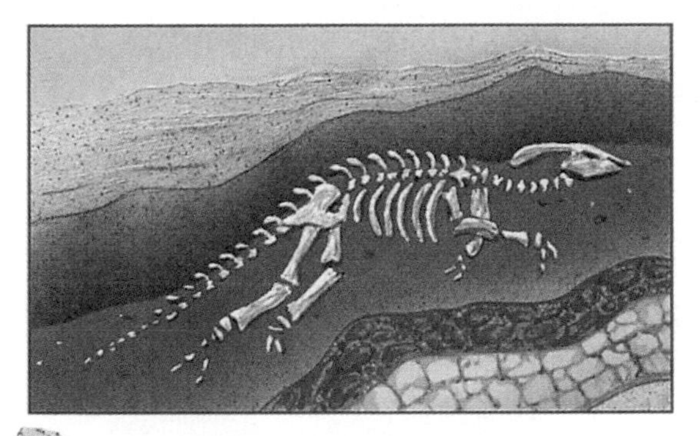

3. The hard parts of the animal or plant—bones, teeth, or woody stems—may harden into fossils.

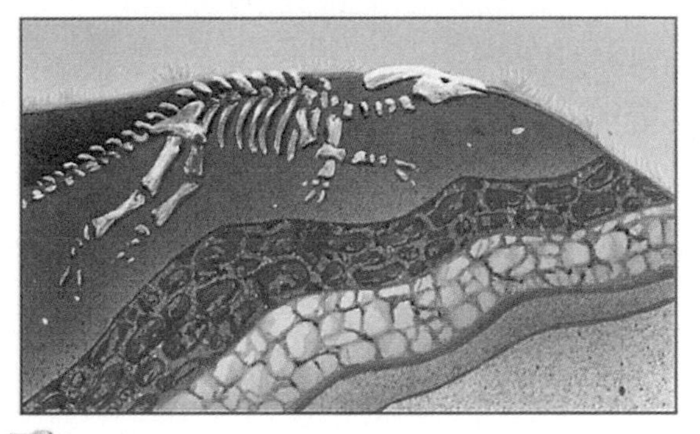

4. Earth changes. Some rock layers get pushed up. Scientists find the buried fossils when they dig in these rock layers.

CHECKPOINT

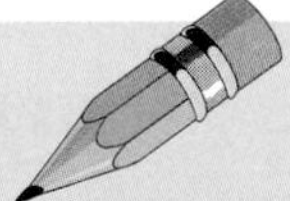

1. Why do many scientists think that woolly mammoths and dinosaurs died out?
2. How do some fossils form?

How can rocks tell you about the past?

ACTIVITY

Making a Fossil

Find Out
Do this activity to learn how to make a fossil model.

Process Skills
Measuring
Constructing Models
Observing
Communicating

WHAT YOU NEED

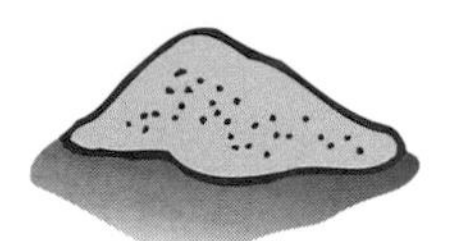

sand

clay

milk carton

rocks

shell

measuring cup

petroleum jelly

goggles

Activity Journal

WHAT TO DO

1. Cover the bottom of the milk carton with sand and rocks.

   Wear goggles when handling sand.

2. Mix sand and rocks into 250 mL of clay.
3. Put half of this mixture on top of the sand and the rocks in the carton.
4. Coat the shell with petroleum jelly and lay it on top of the clay-mix layer.
5. Put the rest of the clay mix on top of the shell.
6. Wait for the clay mix to harden.
7. Tear open the carton and **observe** the layers. Carefully break the layers apart and find your hidden fossil. **Draw** what you see.

What Happened

1. How do your shell and fossil look alike?
2. How do your shell and fossil look different?

What If

What would happen if you used sand instead of clay?

Review

What I Know

Choose the best word for each sentence.

rock	**sand**	**erosion**
weathering	**soil**	**dinosaurs**
crust	**minerals**	**fossils**
climate		

1. A __________ is a piece of Earth's crust.

2. __________ are nonliving material from Earth.

3. __________ is when rocks slowly break apart to make smaller rocks.

4. __________ is made from rocks, minerals, plants, and animals.

5. When rainwater, streams, and wind move or wash soil away, it is called __________.

6. __________ lived on Earth many millions of years before people.

7. __________ buried in rock layers give us clues that tell how animals and plants looked in the past.

Using What I Know

1. How could you describe the rocks in the picture?
2. How will the waves change the shape of the rocks?
3. How else could the rocks be changed?

For My Portfolio

Draw a fossil you would like to find in layers of rock. Tell what plant or animal it comes from.

The SUN, MOON, and EARTH

People have always watched the sun and moon move across the sky. They wondered where the sun goes at night and how the moon changes shape. People all over the world made up stories and legends to explain what they saw. Through the years, people called astronomers studied the sun, moon, planets, and stars. Like the people before them, astronomers saw patterns. Their observations helped them explain how objects in the sky seemed to move.

The Big IDEA

Earth's motion causes night and day and the seasons.

CHAPTER SCIENCE INVESTIGATION

Observe how the moon appears to change shape. Find out how in your ***Activity Journal.***

LESSON 1

Day and Night

Let's Find Out
- What makes a shadow
- How Earth turns

Words to Know

horizon
sunrise
sunset
shadow
axis
rotation

Sun and Shadows

When you look into the distance, you can see a place where the sky meets Earth. This line is called the **horizon.** You cannot see beyond the horizon.

Every morning the sun seems to rise over the horizon. This is called **sunrise.** The sun shines in the sky, and it is daytime.

During the day, the sun seems to move across the sky. It seems to move in the same direction. It seems to move from east to west.

At the end of the day, there is a **sunset.** The sun seems to move below the horizon. The sun seems to set in the west. You cannot see it anymore. Now it is night.

Sunset

During the daytime, you can see your shadow if it is a sunny day. A **shadow** is a place that is dark because something is blocking the light. Your shadow is made because your body blocks the sunlight.

The size of your shadow changes during the day and through the year.

Sundials show how shadows change during the day. People use sundials to tell the time.

Shadows are long when the sun is low in the sky.

Shadows are short when the sun is high in the sky.

Earth Turns

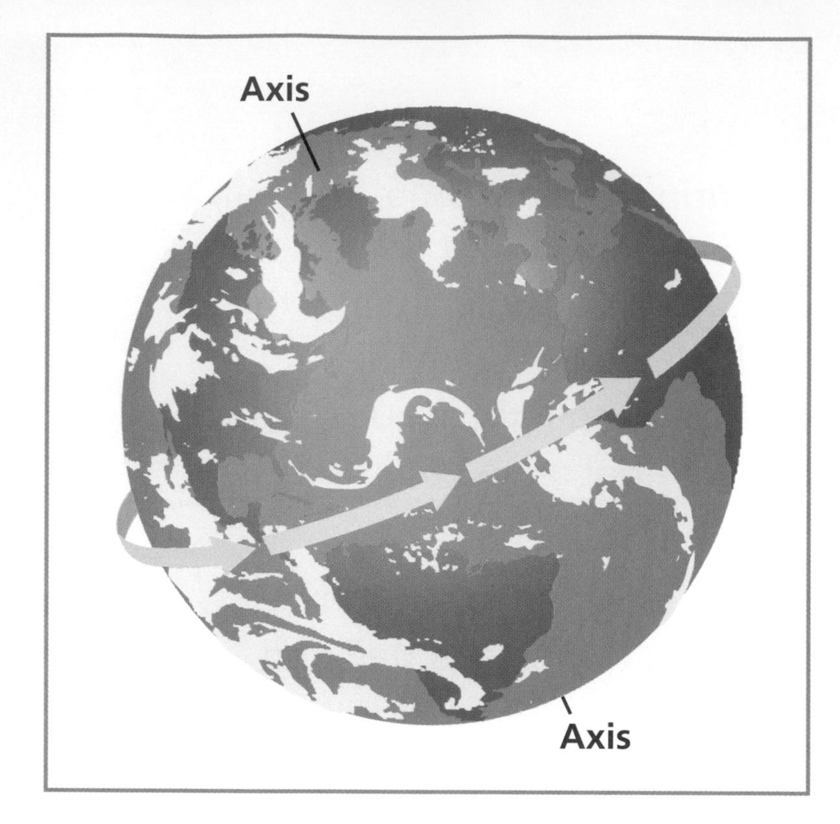

It looks like the sun moves across the sky during the day. If you could stand in outer space, you would see something else. It would look like the sun was not moving.

You would be able to see Earth turning on its axis. The **axis** is an imaginary line that runs through the middle of Earth, from the north pole to the south pole.

Earth is always turning on its axis. Earth's turning is called **rotation.** Earth rotates once each day. It takes 24 hours for Earth to make one complete rotation.

The rotation of Earth causes day and night.

United States

India

Earth is like a giant ball. It is daytime on the part of Earth facing the sun. When it is night where you live, your side of Earth is facing away from the sun. The sun is still shining, but you cannot see it. The sun is shining on the other side of Earth.

India is on the opposite side of Earth from where you live. When you are in school it is night in India. This is because when your side of Earth is facing the sun, India's side of Earth is facing away from the sun.

CHECKPOINT

1. How are shadows made?
2. How does Earth turn?

 What causes day and night?

Activity

Modeling Day and Night

Find Out
Do this activity to see what makes day and night.

Process Skills
- **Constructing Models**
- **Observing**
- **Inferring**

What You Need

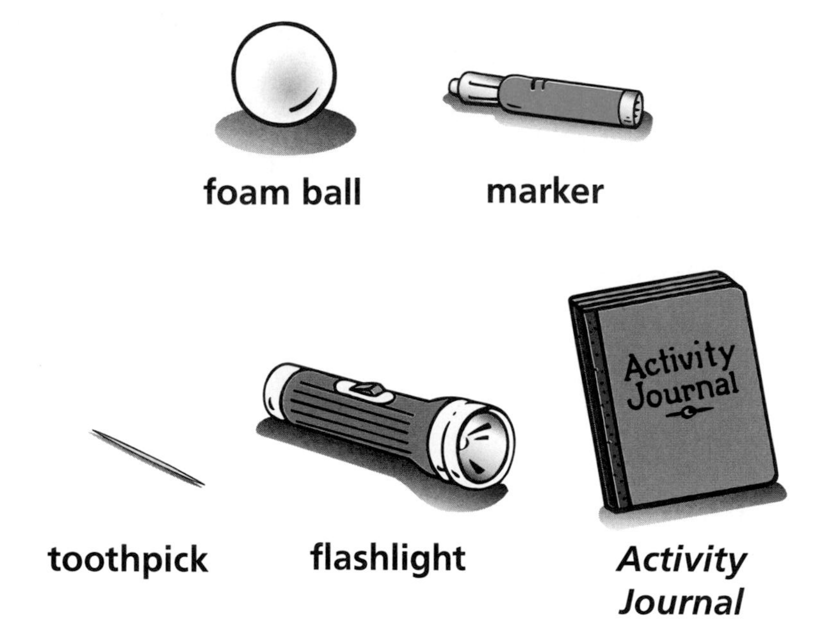

What to Do

1. Find a partner to work with.
2. Stick the toothpick partway into the foam ball.
   Safety! **Be careful with sharp objects.**

3. Use the marker to draw a dot in the middle of the ball.
4. One partner can hold the flashlight. Point the light at the ceiling while the other partner holds the ball by the toothpick. Move the ball so that light shines on the dot.
5. Use the toothpick to spin, or rotate, the ball.
6. **Observe** what happens.

What Happened

1. What object in space is like the ball? What object in space is like the flashlight?
2. What happens when the dot faces away from the flashlight?

What If

What would happen if you did not spin the ball?

Earth and Moon Movement

Let's Find Out

- How Earth moves around the sun
- How the moon moves around Earth
- What moon phases are

Words to Know

revolves
orbit
reflects
phase
new moon
full moon

Earth and Sun

You learned that Earth rotates, but did you know that Earth moves around the sun? Earth **revolves,** or makes a complete trip around the sun. This takes Earth about 365 days. The path that Earth follows around the sun is called an **orbit.** We call one complete orbit around the sun one year.

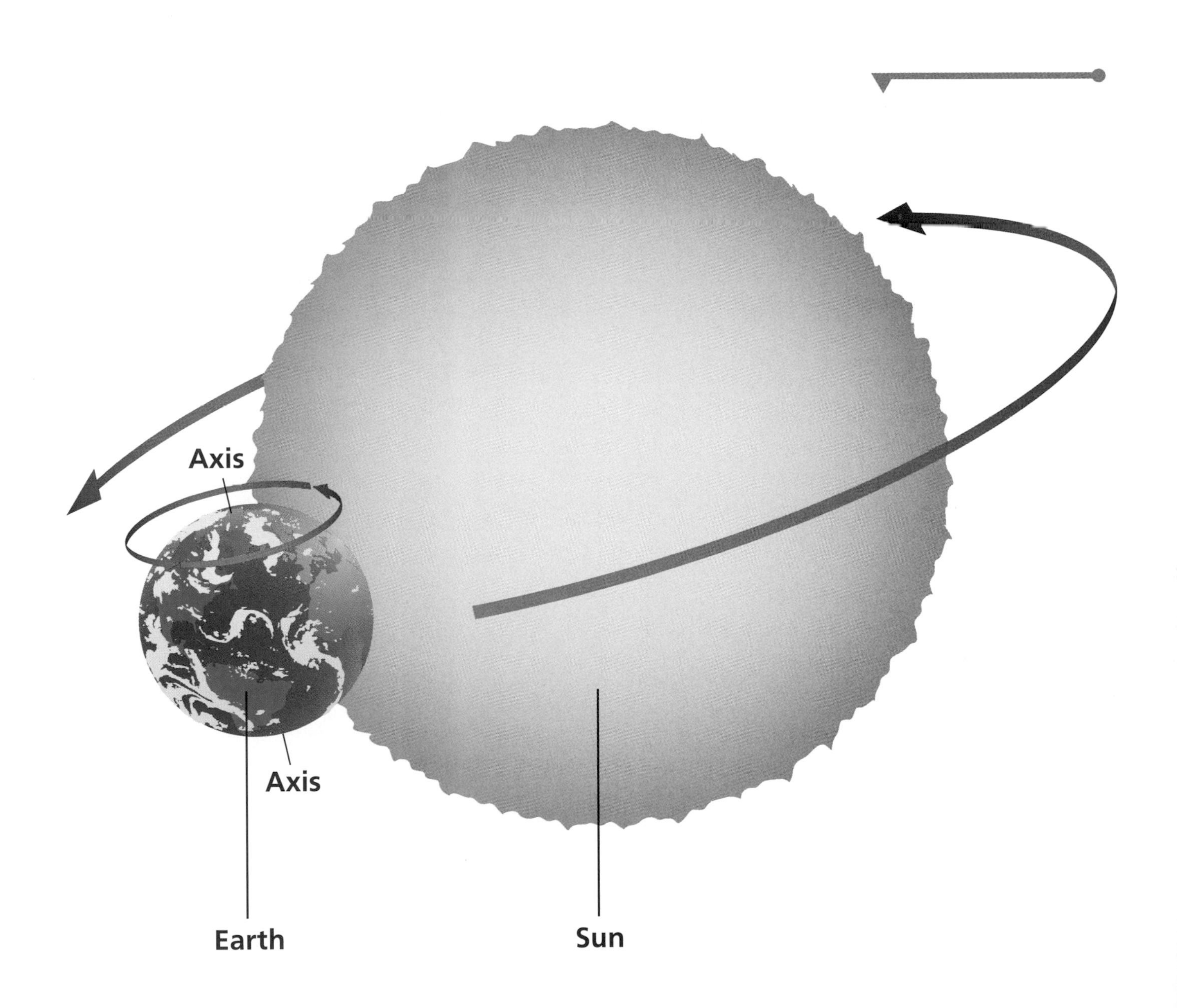
Axis
Axis
Earth
Sun

Earth and Moon

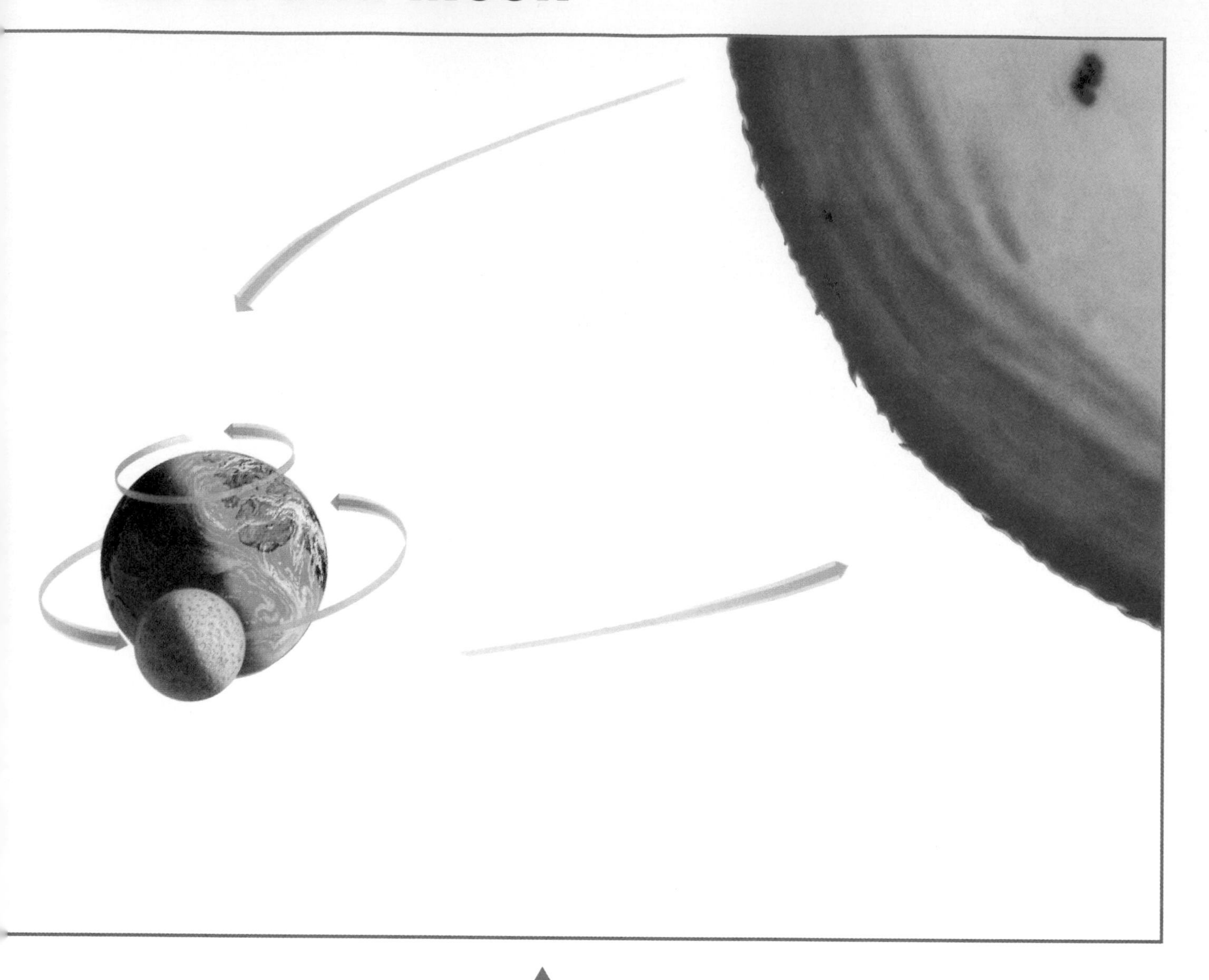

The moon revolves too. The moon moves in an orbit around Earth. It takes the moon about one month to revolve around Earth.

When the moon looks bright at night, people say the moon is shining. But the moon does not make its own light. Light from the sun **reflects,** or bounces, off the moon. It reflects to Earth. Moonlight is reflected sunlight.

From Earth the moon seems to change shape. But the moon always has the same shape. It is shaped like a ball.

Half of the moon is always facing the sun. Sometimes we can see the whole lighted side of the moon. Sometimes we can see only part of the lighted side.

Phases of the Moon

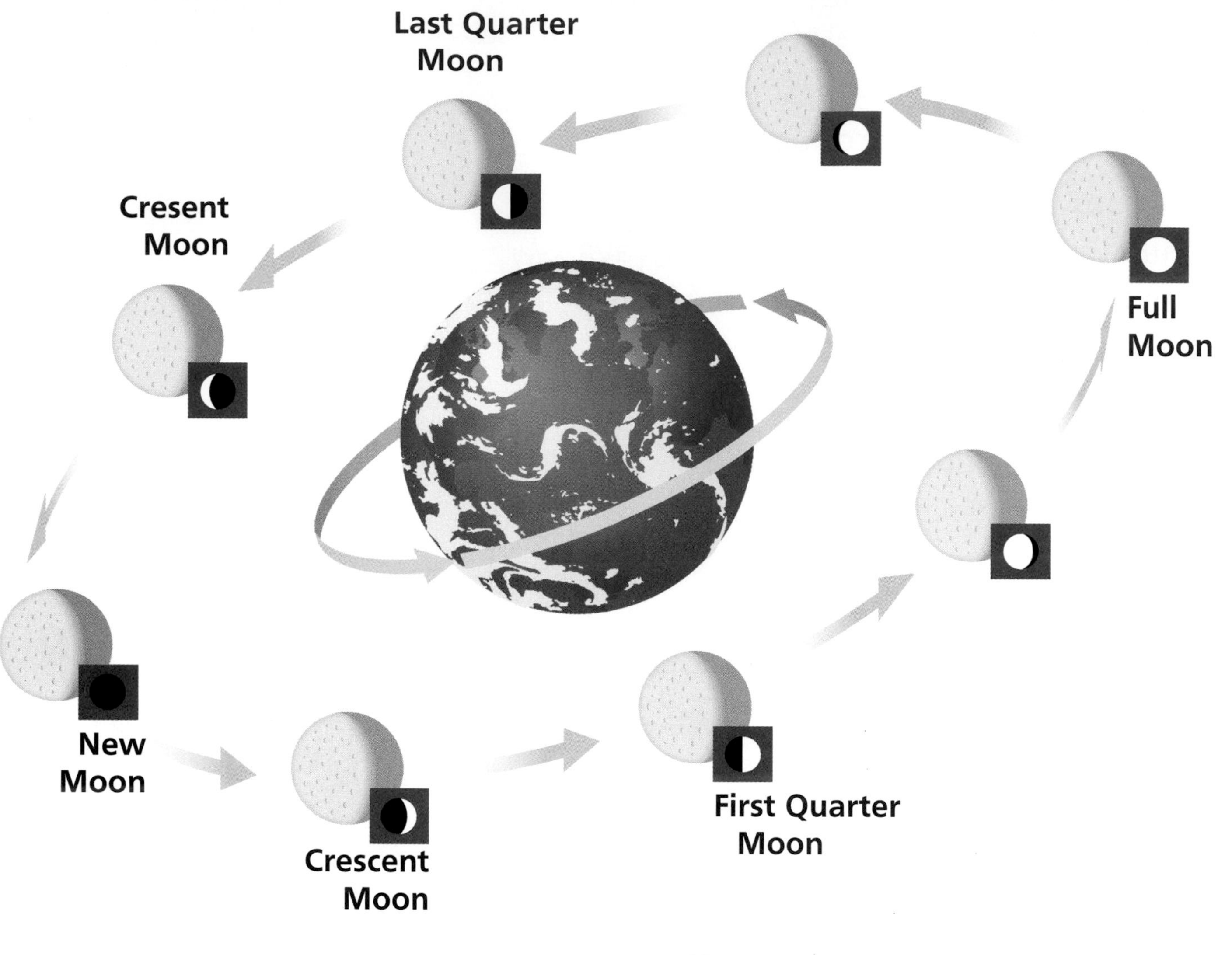

The amount of the lighted side that we can see changes from night to night because the moon is moving. The lighted part of the moon that we see from Earth is called a moon **phase.** Moon phases change in a cycle. The cycle lasts about 29 days.

During the **new moon** phase, all of the lighted side of the moon faces away from Earth. You cannot see the new moon. Each night you see a little bit more of the lighted part of the moon. You see all of the lighted side of the moon during the **full moon** phase. You can see less of the moon each night after the full moon.

CHECKPOINT

1. How long does it take for Earth to revolve around the sun?
2. How long does it take for the moon to revolve around Earth?
3. What are moon phases?

How do Earth and the moon move?

ACTIVITY

Modeling the Moon

Find Out

Do this activity to see why the moon appears to change shape.

Process Skills

- Constructing Models
- Observing
- Communicating
- Predicting

WHAT YOU NEED

foam ball

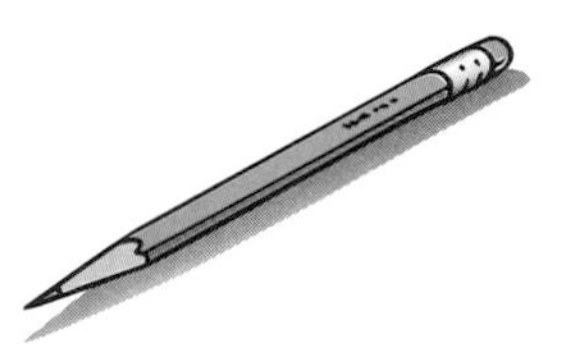

sharpened pencil

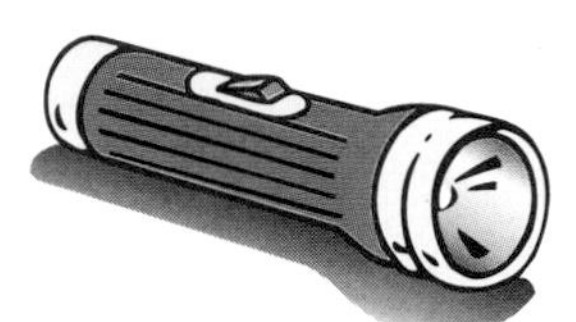

flashlight

Activity Journal

WHAT TO DO

1. Find a partner to work with.
2. Stick the pencil into the foam ball.

   Safety! **Be careful with sharp objects.**
3. In a darkened room, one partner can hold the flashlight

toward the other partner, who holds the ball.

4. Face away from the flashlight and hold the ball in front of you and above your head.
5. Turn around slowly from right to left. Make sure that you keep the ball in front of you and above your head as you turn.
6. As you turn, **observe** the ball.
7. **Record** how the ball changes.

What Happened

1. What object in space is like the ball? What object in space is like the flashlight?
2. What happens to the ball when you are facing toward the flashlight? What happens when you are facing away from the flashlight?

What If

What would happen if you got between the light and the ball?

The Seasons

Let's Find Out

- What happens on Earth as the seasons change
- How places on Earth get different amounts of light and heat

Words to Know

globe
season

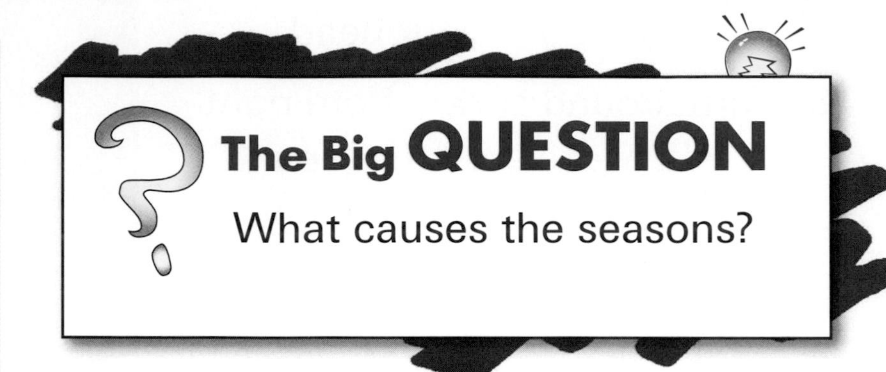

Seasons Change

We can use a globe to learn more about Earth and the sun. A **globe** is a model of Earth. Look at the globe.

You can see that the globe is tilted. Earth is tilted in space too.

As Earth moves around the sun, parts of Earth are tilted toward the sun. These parts of Earth get more light and heat from the sun.

As different parts of Earth get more light and heat from the sun, the seasons change. A **season** is a certain time of year with a particular kind of weather. Seasons change when the amount of light and heat from the sun changes.

When part of Earth is tilted toward the sun the season is summer. It has more daylight hours and it is warmer.

When part of Earth is tilted away from the sun the season is winter. It has fewer daylight hours and it is cooler.

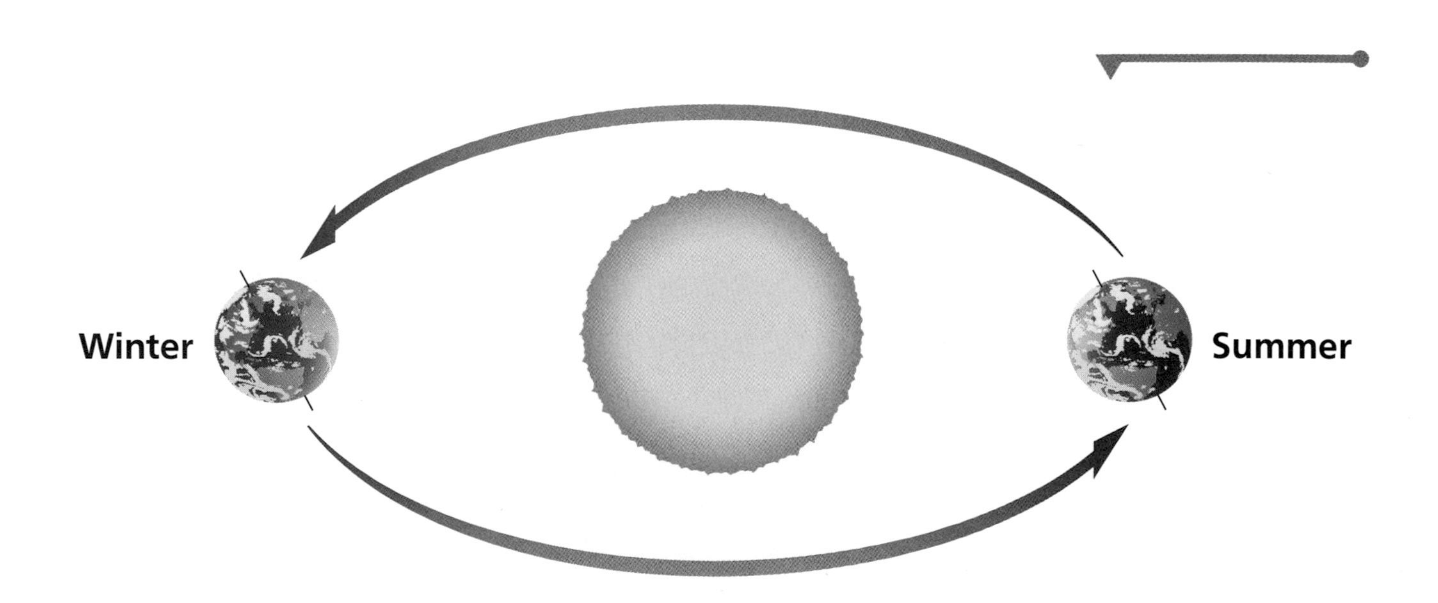

Different Places on Earth

Ecuador is a country near the equator.

Places on Earth do not all get the same amount of light and heat from the sun. Places around the equator, or the middle of Earth, get more light and heat. They are warm all year long. Near the north and south poles, it is cold much of the year. The poles get less light and heat from the sun.

Antarctica is near the south pole.

North America

Australia

You live on the northern half of Earth. Some people live on the southern half of Earth. Seasons are opposite in the two halves. When it is winter in North America, it is summer in Australia.

CHECKPOINT

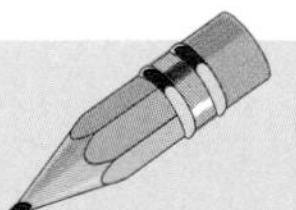

1. What happens on Earth as the seasons change?
2. Where is Earth warm all year long?

What causes the seasons?

ACTIVITY

Shining Sunlight

Find Out
Do this activity to see how summer and winter are different.

Process Skills
Constructing Models
Using Numbers
Communicating
Observing
Inferring

WHAT YOU NEED

flashlight clay graph paper

Activity Journal

5-cm piece of straw

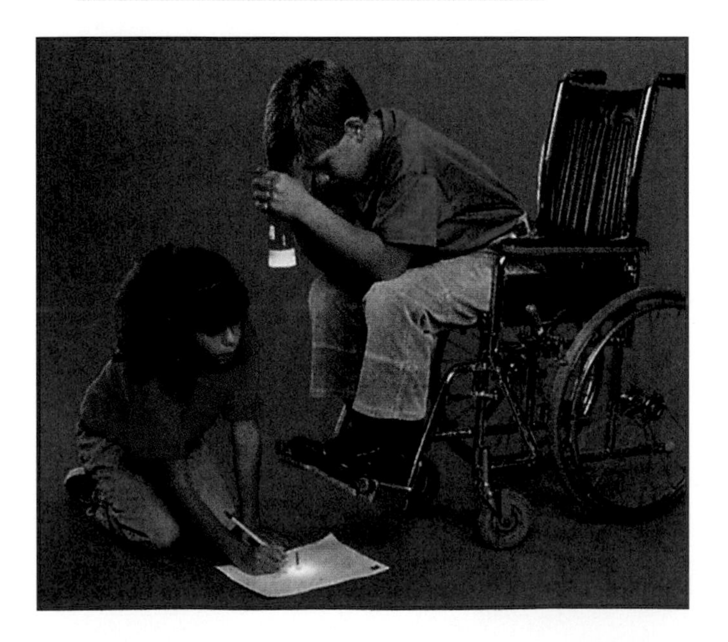

WHAT TO DO

1. Put a piece of graph paper on the floor. Place clay in the middle of the paper. Poke the straw straight into the clay.
2. Work in a darkened room. Sit in a chair and rest your elbows on your legs, holding the flashlight in your hands.

3. Shine the flashlight straight down above the straw.
4. Have a classmate trace the outline of the lighted area on the paper. **Count** the squares inside the outlined area.
5. Now, while staying in the same place, hold the flashlight so it is tilted. Move the paper so the clay is at the center of the light. Be sure the distance from the flashlight to the paper is the same as before. **Record** what you **observe**.
6. Trace the outline of the lighted area and **count** the squares.

WHAT HAPPENED

1. Imagine the flashlight is the sun and the light on the paper is sunlight as it hits Earth. Which try is like the sun and Earth in winter? Which try is like summer?
2. Why did the outlines change?

WHAT IF

What would happen if the flashlight were always held straight down?

Review

What I Know

Choose the best word for each sentence.

phase	**revolves**	**sunrise**	**axis**
globe	**full moon**	**horizon**	**sunset**
shadow	**rotation**	**season**	**orbit**
	new moon	**reflects**	

1. The place in the distance where the sky and Earth meet is called the __________.

2. A __________ is a place that is dark because something is blocking the light.

3. Earth's __________ is an imaginary line that runs through the middle of Earth from the north pole to the south pole.

4. Earth's turning is called __________.

5. Earth __________ around the sun once about every 365 days.

6. The lighted part of the moon that we see from Earth is called a moon __________.

7. A __________ is a model of Earth.

Using What I Know

1. What causes day and night?

2. What causes moonlight?

3. Why does the moon seem to change shape?

For My Portfolio

What season is it right now? Draw a picture of what your neighborhood looks like during this season.

Unit Review

Telling About What I Learned

1. People use Earth's resources in many ways. Name four natural resources.

2. Rocks, minerals, and soil make up Earth's land surface. Tell what makes up soil.

3. Earth's motion causes night and day and the seasons. Name two ways Earth moves.

Problem Solving

Use the picture to help answer the questions.

1. What natural resources can you see?

2. What could cause the rocks in the picture to change over time?

Something to Do

Imagine a moon town. Make a list of things you would need to bring from Earth to your moon town. What resources are they made of? Make a mural showing how your moon town looks and how you get the things you need to survive.

UNIT C

Physical Science

CHAPTER 1

MOTION

Children walk. Dogs run. A car drives by. Paper blows. A ball flies through the air.

Animals use muscles to move. A motor moves a car. Wind blows the paper. A person throws the ball.

The Big IDEA

Objects move in ways that may be predictable.

CHAPTER SCIENCE INVESTIGATION
Make a boat and see how the size of the load affects how it moves. Find out how in your *Activity Journal.*

LESSON 1

Movement

Let's Find Out

- How people use pushes and pulls
- What causes things to fall down

Words to Know

position
motion
forces
gravity

The Big QUESTION

What causes objects to move?

Pushes and Pulls

You can talk about where things are. A book can be on a desk. A baseball can be in the air. The **position** of something is the place where it is. You know your position by looking at the objects around you. You can tell how far away you are from another object.

People and objects can change position. **Motion** happens when something changes position. Throwing a baseball to a friend changes the position of the ball. The ball moves when you throw it.

The ball does not start moving by itself. Your arm pushes the ball. Pushes and pulls change the way objects move. All pushes and pulls are **forces.**

You start the day by pulling or pushing yourself out of bed. All day long, you push doors open and you pull them closed. Without forces you couldn't live as you do now. How are these people using forces?

Gravity

When you throw a ball, it lands somewhere. It does not fly off into space. It falls to the ground. It falls because of the force of **gravity.**

The force of gravity pulls objects down to Earth. Earth's gravity is always pulling things toward Earth.

Any time you pick something up, you are using a force. You are using force to work against gravity. When you jump in the air, your body comes down to the ground. Gravity pulls you back down. Without gravity you would fly off into space.

Gravity makes the water in this stream fall.

CHECKPOINT

1. How do people use pushes and pulls?
2. What force causes things to fall down?

What causes objects to move?

ACTIVITY

Observing Movement

Find Out
Do this activity to learn how pushes move objects.

Process Skills
- Observing
- Measuring
- Communicating
- Predicting
- Interpreting Data

WHAT YOU NEED

tape

table-tennis ball

straws

meterstick

Activity Journal

WHAT TO DO

1. Make a mark on the ground with a piece of tape. Put a table-tennis ball on the mark.
2. Blow one puff of air toward the ball. **Observe** the ball move.

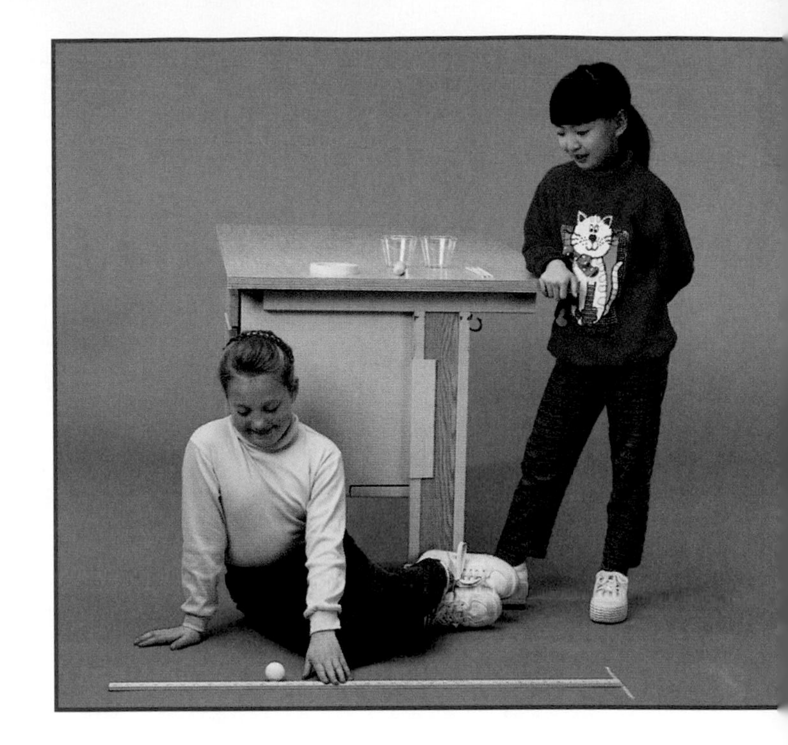

3. **Measure** the distance from the mark to the ball. **Record** what you find.
4. Repeat steps 2 and 3 three more times.
5. **Predict** how far the ball will move if you blow on the ball and your partner blows on the ball from the opposite direction.
6. **Measure** and **record** how far the ball moves.
7. Repeat steps 5 and 6 three more times.
8. **Make a graph** of your measurements.

What Happened

1. How far did the ball move each time?
2. What pushed on the ball to make it move?

What If

What other forces could move the ball?

Force and Motion

Let's Find Out

- Why some objects move faster than others
- What force makes an object slow down

Words to Know

speed
direction
friction

Fast and Slow

When you kick a soccer ball, how fast does it go? How fast an object moves is its **speed.** You can measure the speed of the ball by timing how fast it moves from one position to another. The ball's speed depends on how heavy it is, how strong the push or pull is, and how long the push or pull lasts. If you want to change how fast a ball is moving, you can change the amount of force that you use. You can give the ball a big push all at once or you can lightly push the ball for a long time.

Which way an object moves is called its **direction** of motion. You can throw a paper plane across the room or out the window. You can choose which direction to throw it.

The direction of moving objects can change. You can push a moving soccer ball with your feet to change its direction. Forces can change an object's speed, direction, or both.

A car slows down to stop at a red traffic light. The light turns green. The car speeds up again. Then it turns onto another street. Speeding up, slowing down, and turning all happen because of forces.

Slowing Down

When you roll a ball on the ground, it does not keep rolling forever. The ball rubs against the ground and slows down. Finally, it stops. The force that stops the ball is **friction.** Friction is caused by two things rubbing against each other. Friction can slow an object down.

If you walk around on a wooden floor wearing socks without shoes, you might slide around on the floor. If you wear tennis shoes you will not slide around very much. The rough bottoms of the shoes rub against the floor. Friction helps you not to slide.

Many slides are made of plastic because plastic surfaces do not cause a lot of friction.

Anything that touches an object causes friction. Even air and water cause friction. Water slows down a moving boat. When you push a spoon through a bowl of soup, friction from the soup slows the spoon.

You use friction every day. Without friction you could not walk. Your feet push against the ground when you walk. Friction between your feet and the ground helps you walk.

CHECKPOINT

1. Why do some objects move faster than others?
2. What force makes moving objects slow down?

 How can an object's motion change?

Activity

Investigating Friction

Find Out

Do this activity to see how friction changes the movement of objects.

Process Skills

Observing
Communicating
Inferring

What You Need

sandpaper

four thumbtacks

smooth board

three wooden blocks

masking tape

Activity Journal

What to Do

1. Tape the sandpaper to one side of a block, so the rough side faces out.
2. Push the tacks into the four corners of one side of another block.

   Safety! **Use tacks carefully.**
3. Leave the third block alone.

4. Line up all three blocks on one end of the board. The blocks with sandpaper or tacks should have those sides down.
5. Raise the end of the wood to make a ramp. **Observe** the blocks as they move.
6. **Record** what happened.

What Happened

1. Which block moved first? Which block moved last?
2. Why didn't all of the blocks move as soon as you started to raise the board?

What If

How could you change the blocks or the ramp to create less friction?

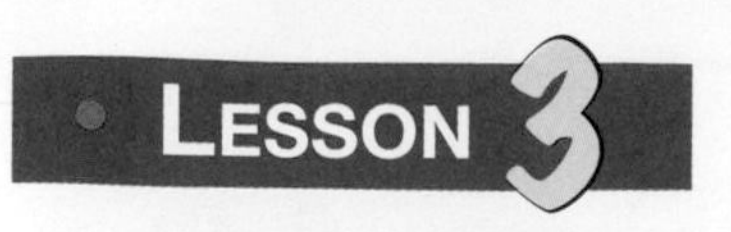

Machines and Magnets

Let's Find Out
- How machines help people
- How magnets work

Words to Know
machines
magnets
poles

The Big QUESTION
How can machines and magnets change an object's motion?

Machines

People use their muscles to move objects. People also use machines to move objects. **Machines** are tools used to apply a force. Machines help people pull or push objects. Machines come in all sizes. A machine can be smaller than a toothpick or larger than a crane.

You use machines every day. A hammer is a machine that can push a nail into wood. It can also pull the nail out of the wood.

Some bottles are hard to open. You can't pull off the top with the force of your fingers. Using a bottle opener increases your force. The force of the bottle opener pulls off the bottle top.

Think of how you sharpen your pencil. A pencil sharpener is a machine that helps you.

Think of how you eat your food. A knife helps you cut your food. A knife is a machine.

Magnets

Magnets can push and pull objects too. Magnets attract and repel each other. This means they pull each other together and push each other apart. Magnets can move some objects without ever touching them. Magnets attract some metal objects.

The force of a magnet is strongest at its ends. The ends of a magnet are called **poles.** One end is called the north pole. The other end is called the south pole. The north pole of one magnet attracts the south pole of another magnet. Opposite poles pull toward each other. Same poles push each other away. A north pole repels another north pole.

This crane uses a magnet to pick up scrap metal.

Objects can stay in one place because of forces like gravity and friction. Forces can also make objects move or change their direction of motion. Machines and magnets can make objects move. They push or pull against gravity and friction.

CHECKPOINT

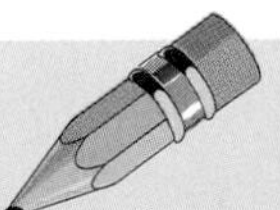

1. How do machines help people?
2. Which ends of magnets are attracted to each other?

How can machines and magnets change an object's motion?

ACTIVITY

Observing Magnets

Find Out
Do this activity to see how magnets pull and push each other.

Process Skills
Observing
Communicating
Inferring

WHAT YOU NEED

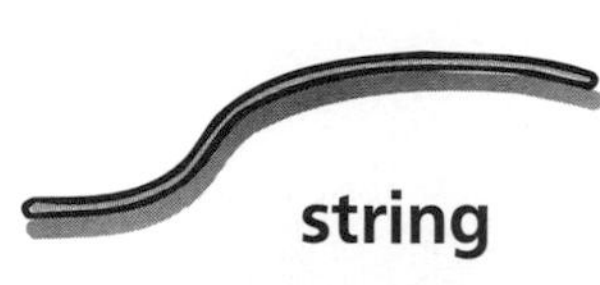
string

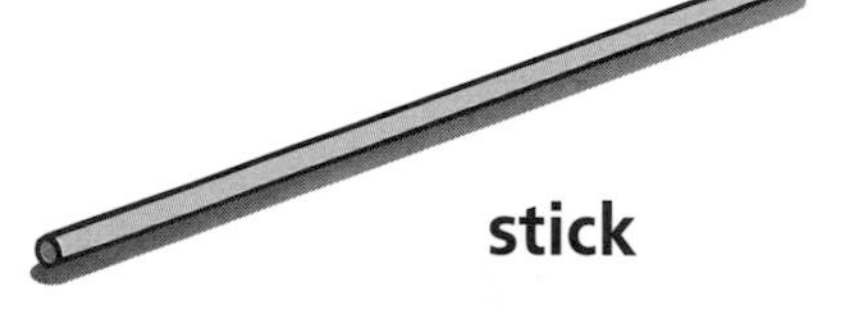
stick

Activity Journal

two bar magnets

two chairs

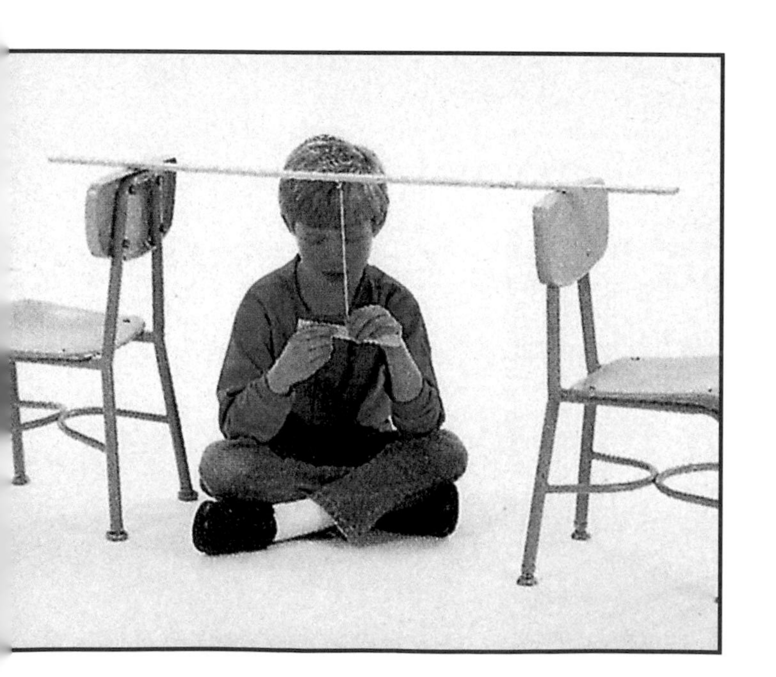

WHAT TO DO

1. Balance the stick across the chairs. Tie the string to one magnet. Hang the magnet from the stick.

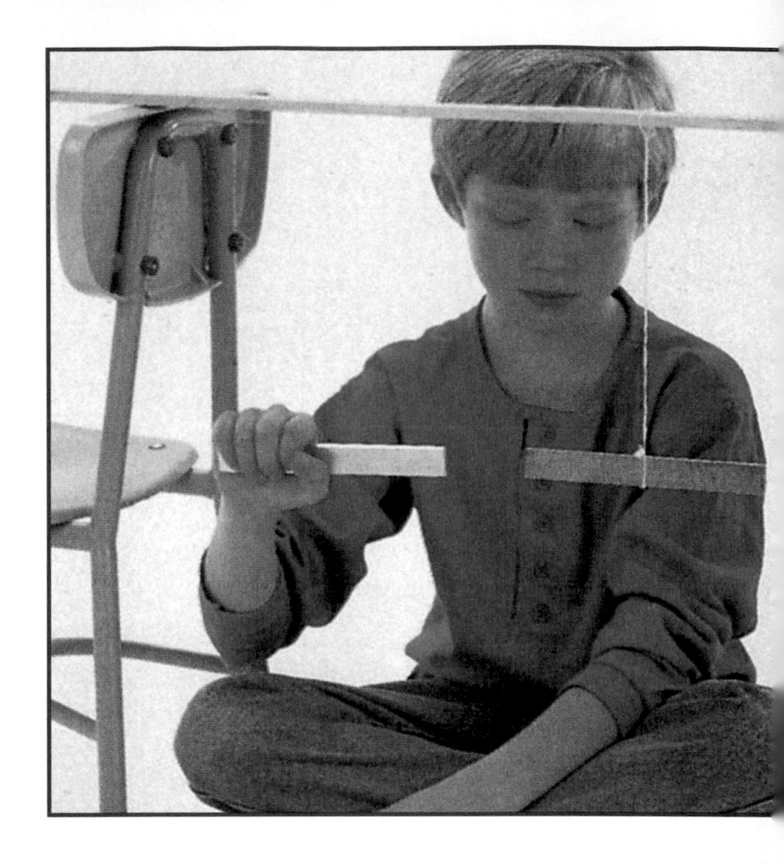

2. Hold the other magnet in your hand. Hold one end of the magnet up to each end of the hanging magnet.
3. **Observe** what happens. **Record** what you see.
4. Hold the other end of the magnet to each end of the hanging magnet. **Observe** what happens. **Record** what you see.

What Happened

1. Which ends of the magnet were attracted? Which ends were not attracted?
2. What happened when the ends were not attracted?

What If

How could you use these magnets to help you do a job?

Review

What I Know

Pick the best word for each sentence.

position	**direction**	**motion**
friction	**forces**	**machines**
gravity	**magnets**	**speed**
poles		

1. All pushes and pulls are __________.

2. __________ pulls objects down to Earth.

3. __________ is the measure of how fast an object is moving.

4. Which way an object moves is called its __________ of motion.

5. A force that causes objects to slow down is called __________.

Using What I Know

1. What is the girl in this picture doing?
2. How is the tool helping her?
3. What makes the nail move?
4. What will happen if something falls off the table?
5. What is the magnet doing to the other nails?

For My Portfolio

Find a few different types of balls such as soccer balls, kick balls, and baseballs. Take them outside and see how far you can kick each type of ball. Describe why each ball moved as it did.

Sounds are all around. People whisper and laugh. A computer hums. A door creaks. A car horn honks. Even when it seems quiet, there is always something to hear.

What sounds does this picture make you imagine?

The Big IDEA

Sound can be described by its volume and pitch.

CHAPTER SCIENCE INVESTIGATION

Make musical instruments and play them. Find out how in your ***Activity Journal.***

LESSON 1

How Sound Is Made

Let's Find Out

- How sound moves
- How some animals make sound

Words to Know

vibrate
vibrations
sound waves

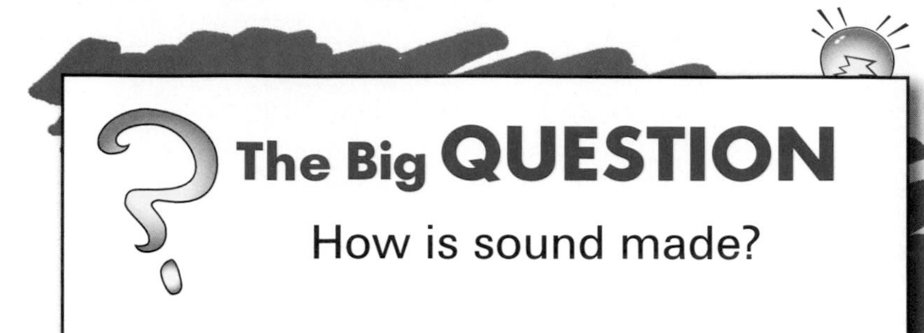

Sound Waves

Have you ever heard the sound of crashing cymbals? That sound forms when one cymbal strikes the other. When the cymbals bang together, they **vibrate.** They move back and forth. The **vibrations** move the air around them back and forth. The vibrations make **sound waves** in the air. The sound waves spread through the air. You cannot see sound waves,

but they are there. If you could see them, you would see them move like water waves.

The sound waves move through the air until they reach the eardrum inside the ear. The vibrating sound waves cause the eardrum to vibrate too. It vibrates in the same way as the cymbals.

Making Sound

You can make sounds by hitting things or tapping objects. You can make sounds by plucking strings or snapping your fingers.

The woman in the picture is playing a guitar that has six strings. When the strings are plucked or strummed, they vibrate. The vibrations make sound. The drummers make sounds by hitting their drums with wooden sticks and mallets.

Some things vibrate so quickly you cannot see them move. Bumblebees move their wings very fast. The wings vibrate. The vibrations move air. You hear a buzzing sound.

Male crickets rub their wings back and forth. The vibration makes a chirping sound.

People talking make sounds with vocal cords in their throats. The vocal cords are like strings that vibrate. They vibrate as air from the lungs moves over them.

CHECKPOINT

1. How does sound move?
2. How do people and some other animals make sound?

 How is sound made?

ACTIVITY

Making Sound With a Drum

Find Out
Do this activity to see and hear sound waves.

Process Skills
Observing
Communicating
Inferring

WHAT YOU NEED

pencil with eraser

small coffee can

thick rubber band

big round balloon

Activity Journal

goggles

scissors

WHAT TO DO

1. Use scissors to cut off the small end of a balloon.

   Be careful with scissors.

2. Stretch the balloon over the open end of the coffee can. Be careful so the balloon does not break.
3. Put the rubber band around the can to hold the balloon in place.

   Safety! **Be careful with rubber bands.**
4. Tap your drum with the eraser end of the pencil. **Observe** what happens.
5. **Record** what you see.

What Happened

1. How do you make a sound with your drum?
2. What part of the drum vibrates?

What If

What are other ways that you could make sound with your drum?

Volume

Let's Find Out
- Why some sounds are louder than others
- How you can make sounds louder or softer

Words to Know
loud
soft
volume
noise

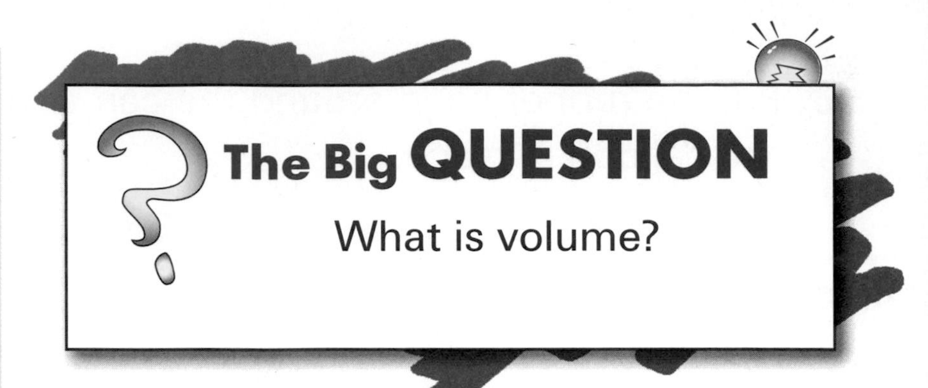

Loud and Soft

Think of a thunderstorm. You hear quiet raindrops on the roof. Then, you hear a **loud** thunderclap. Why are some sounds so loud and other sounds so **soft?**

Bigger forces make louder sounds. A big force makes a thunderclap. Small forces make softer sounds. That is why raindrops make softer sounds.

Thunder makes a loud sound.

Volume is the loudness or softness of a sound. A loud sound has a high volume. A soft sound has a low volume.

You can change the volume of your voice. You can speak softly. You can shout loudly. Shouting takes more force than talking quietly.

More force makes stronger vibrations. Stronger vibrations make louder sounds. Rub your hands together. Snap your fingers. Now, clap your hands. The sound is loudest when you clap your hands, because clapping your hands uses more force than rubbing your hands together or snapping your fingers.

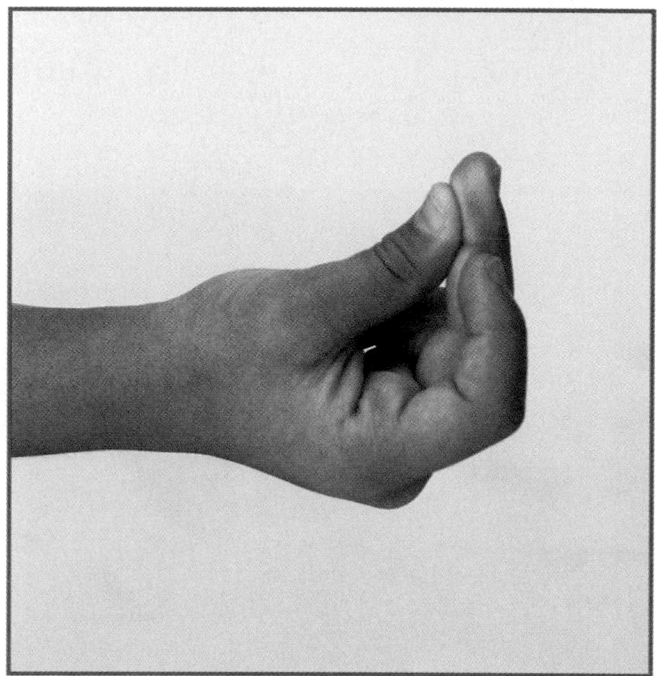

Louder and Softer

You can make sound louder or softer. This man uses a megaphone to collect sound. His megaphone sends all of the sound waves in one direction. This makes the sound easier to hear.

Listen to sounds around you. Now, cup your hands around your ears. This collects the sound. You can hear quieter sounds.

Loud sounds can be dangerous. Loud sounds can hurt your ears.

There are ways to make sounds softer. You can turn down the volume on the radio.

You can cover something that is loud. Cover an alarm clock with a blanket. The sound will be less loud. You can also cover your ears.

Some sounds are pleasant, but some are not. Unpleasant sounds are called **noise.**

People who work in very noisy places protect their ears. They cover their ears. Then the noise does not sound as loud.

Ears need to be protected from loud noise.

CHECKPOINT

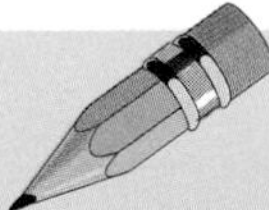

1. Why are some sounds louder than others?
2. How can you make sounds louder or softer?

What is volume?

ACTIVITY

Changing Volume

Find Out

Do this activity to change the volume of sounds.

Process Skills

- Observing
- Communicating
- Predicting
- Inferring

WHAT YOU NEED

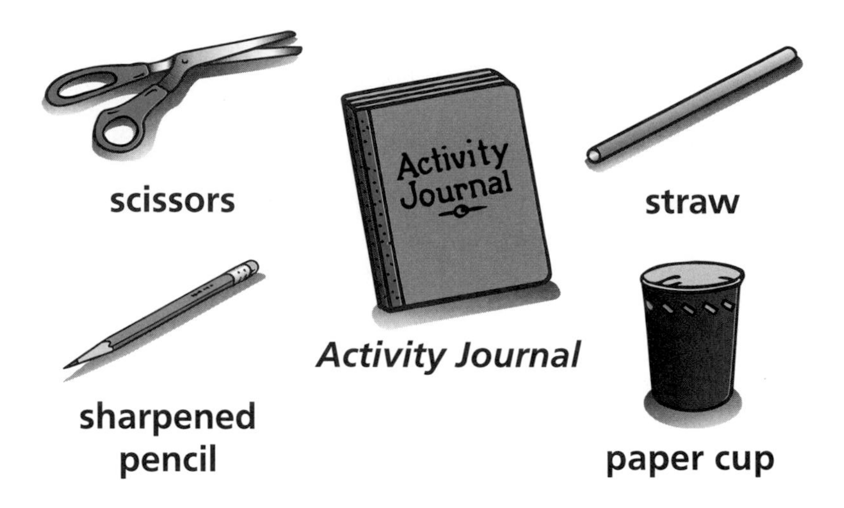

WHAT TO DO

1. Make a straw horn. Flatten the straw by sliding your fingers up and down the length of it.
2. Use your scissors to cut the end of the straw to make a shape like an arrow.

   Safety! **Be careful with scissors.**
3. Blow into the pointed end of the horn to make a sound.

4. Use your pencil to poke a hole in the bottom of the cup. Safety! **Be careful poking the hole.**
5. Put a cup on the straight end of the horn.
6. Blow into the horn again. **Observe** how the sound changes.
7. **Record** what happened.
8. **Predict** what would happen if you covered the cup with your hand while you blew into the straw.

What Happened

1. Which time was the sound loudest?
2. What caused the sound to be louder?

What If

How could you make the sound even louder?

Pitch

Let's Find Out
- Why some sounds are high and others are low
- How musicians use pitch

Words to Know
pitch
music
musical sounds

High and Low

When you sing a song, some parts have high sounds. Some parts have low sounds. High and low describe the **pitch** of the sound. Every sound has a pitch. A big dog's bark has a low pitch. A small dog's bark has a high pitch.

You learned that vibrations make sounds. Things that vibrate slowly make low sounds. Things that vibrate quickly make high sounds.

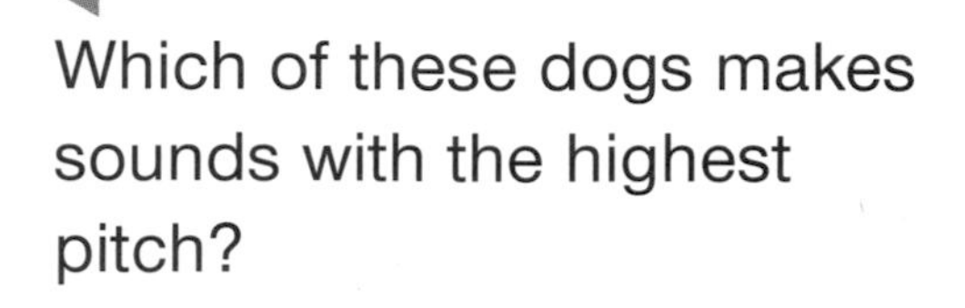
Which of these dogs makes sounds with the highest pitch?

Small bells vibrate faster than larger bells. They make higher pitches. Larger bells make lower pitches.

Guitar strings make different pitches. Short, thin strings vibrate faster. They make higher pitches. Long, heavy strings vibrate slower. They make lower pitches. The guitar player can also make a string sound higher. The player presses a string down to make it shorter.

Using Pitch

Pianos can play many pitches. They can play low notes or high notes. Piano players can play **music** on a piano. They press the keys to make **musical sounds.**

Music has a pattern of pitches. We use musical notes to show pitch. Musicians read the notes and play the correct pitches on their instruments.

A choir is made up of many people with different voices.

You can describe a sound by its pitch and volume. But that does not tell you everything about a sound. Guitars and trumpets can play sounds of the same pitch, but they sound different. Think about it. When you sing a song with your friend, you both sing the same pitches. But, your voice sounds different from your friend's voice. There are many ways to describe sounds.

CHECKPOINT

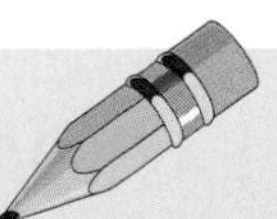

1. Why are some sounds high and others low?
2. How do musicians use pitch?

 What is pitch?

ACTIVITY

Making a Bottle Instrument

Find Out
Do this activity to experiment with pitch.

Process Skills
- Measuring
- Observing
- Predicting
- Classifying
- Communicating

WHAT YOU NEED

water

wooden spoon

six clear, empty bottles

Activity Journal

WHAT TO DO

1. **Measure** and pour a different volume of water into each bottle.
2. Tap two of the bottles. **Observe** the sounds.

3. **Predict** which bottle will make the highest sound when the bottles are tapped.
4. Carefully tap each bottle with the wooden spoon.
5. **Observe** what happens.
6. Put the bottles in order from the lowest water level to the highest.
7. Tap them again. **Record** what happens.

What Happened

1. What did you notice about the amount of water in each bottle and the noise it made?
2. What is vibrating to make the sounds?

What If

What would happen to the sounds if you added more water to each bottle?

Review

What I Know

Pick the best word for each sentence.

sound waves	**vibrations**	**vibrate**
loud	**soft**	**volume**
pitch	**music**	**noise**
musical sounds		

1. ___________ are back-and-forth movements.

2. Big forces make ___________ sounds.

3. The loudness or softness of a sound is its ___________.

4. A fast vibration makes a high ___________.

5. ___________ has a pattern of pitches.

Using What I Know

1. What musical instruments do you see in the picture?
2. How do the different instruments make sound?
3. How can the drummer make a louder sound?
4. How can the trumpet player play different pitches on the trumpet?
5. How can the banjo player change the pitch of the banjo?

For My Portfolio

Make up a song about how sound is made. Try singing louder and softer. Try singing higher and lower.

LIGHT AND HEAT

Living things need the sun. Even though the sun is far from Earth, it gives Earth light and heat. Light from the sun moves through outer space. The light shines on Earth. Sunshine warms Earth. Living things need light and heat from the sun.

The Big IDEA

Light and heat can travel and be absorbed.

CHAPTER
SCIENCE
INVESTIGATION
See what happens to
light when it shines on
different materials.
Find out how in your
Activity Journal.

Lesson 1

Light

Let's Find Out
- How light helps people see
- What happens when light shines on objects

Words to Know
- **light**
- **reflects**
- **absorb**
- **shadow**

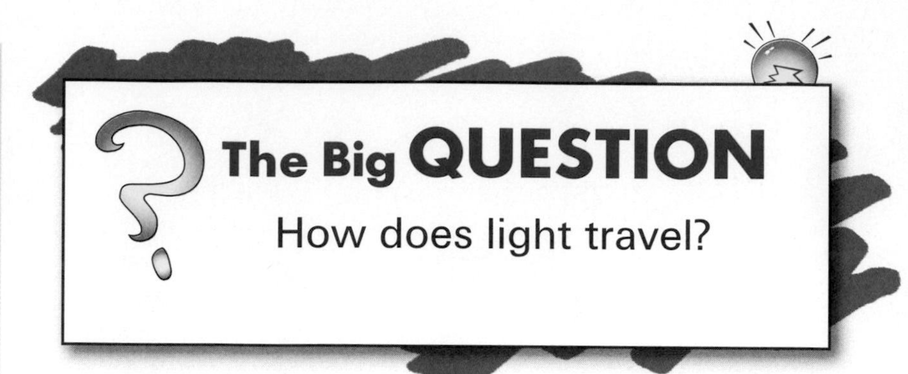

Light and Seeing

Light helps us see. Most objects do not make their own light. You can only see them when light shines on them.

It is bright when there is a lot of light. In bright light you might see many colors. Name some colors you can see. When there is less light it is dim. Colors are harder to see in dim light. When there is no light it is dark. Everything looks black.

Light moves in straight lines. When it hits a surface, some light bounces off. The surface **reflects** the light. The reflected light strikes your eyes. You see the object.

Not all of the light gets reflected. A surface might **absorb** some of the light. The light it absorbs warms the object.

Dark-colored surfaces absorb more light. They reflect less light. Light-colored surfaces absorb less light. They reflect more light. Some surfaces reflect almost all the light that hits them. Bright and shiny surfaces reflect the most light.

Which team probably feels hotter?

Mirrors reflect a lot of light.

Light Shines

Some materials let light pass through. Glass and water let a lot of light through. That is why you can see through glass and water.

Other materials let some light through. You can see light behind them. But you cannot tell what else is behind them.

When light passes through some things, it looks different. The light changes color.

Light can pass through glass.

Shadows form when light shines on something but does not pass through it.

Could you see a light through a wooden door? No, because light travels in straight lines. Some materials block light. So light cannot get through them. If light hits a material that it cannot pass through, a **shadow** can form.

Shadows can be different sizes. Sunlight makes shadows. The shadows change through the day.

CHECKPOINT

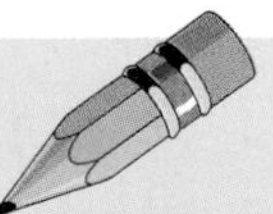

1. How does light help people see?
2. What happens when light shines on objects?

How does light travel?

ACTIVITY

Investigating Light

Find Out
Do this activity to discover how light travels.

Process Skills
- Measuring
- Observing
- Predicting
- Inferring

WHAT YOU NEED

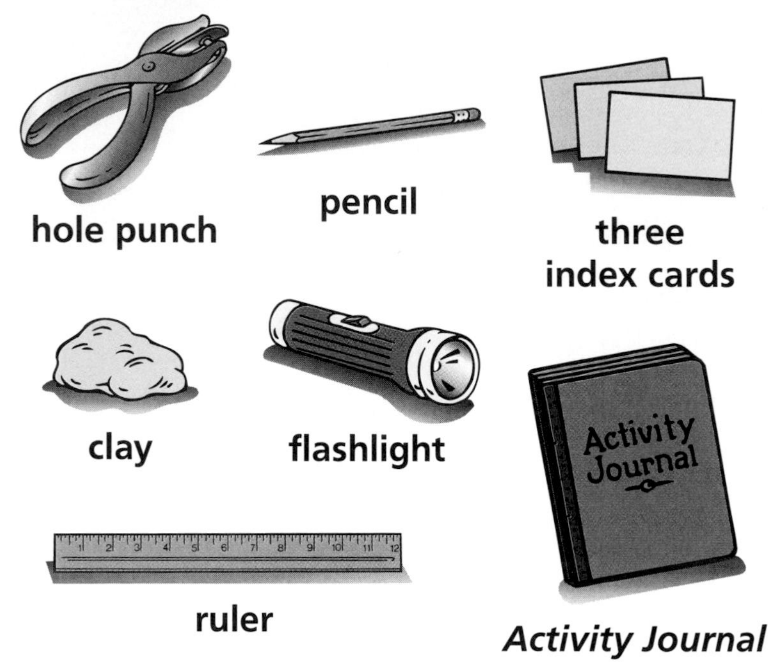

WHAT TO DO

1. Use the ruler to draw lines on each card from one corner to another to make an X.
2. Punch a hole in each card where the lines cross. Make six small balls of clay.

3. Stand the cards up in the balls of clay. **Measure** 15 cm between each card. The cards should be behind each other, so that you can see straight through the holes.

4. Ask a partner to shine a flashlight straight into the first hole. **Observe** the light.

5. Have your partner shine the light into the hole from the side. **Observe** what happens.

6. Move the cards so that you cannot see straight through the holes. **Predict** what will happen. Try the light again and **observe** what happens.

What Happened

1. What changed when you moved the cards?
2. How did you make the light travel through the holes in all three cards at the same time?

What If

Would you see any light if you shined the flashlight on index cards that did not have holes?

Heat

Let's Find Out

- How people use heat
- How people can make heat without using fuels

Words to Know

heat
fuel
electricity
friction

The Big QUESTION

Where does heat come from?

Using Heat

People use **heat** in many ways. A burning object makes heat. A material that is burned to make heat is called a **fuel.**

People use different fuels to make heat. Wood, natural gas, and oil are fuels. Some people burn wood in a stove or fireplace to make heat. You may burn natural gas in a stove in your kitchen. A car's engine burns gasoline. Gasoline is made from oil. Some buildings are heated by burning coal. Coal is a fuel, too.

Heat from coal, oil, and natural gas can be used to make **electricity.** Electricity can then be used to heat a home. How is your home heated? How do we use heat in the objects in the pictures?

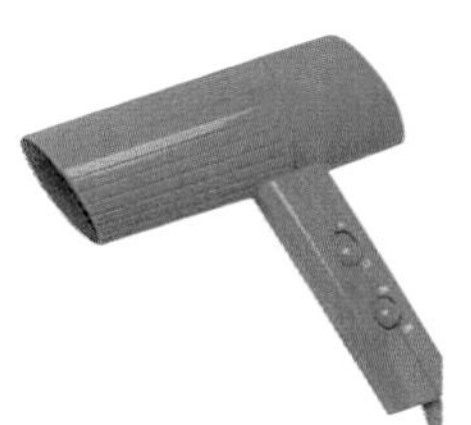

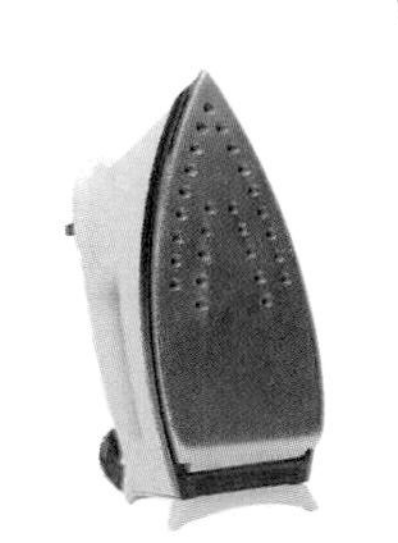

When the weather is cold, some people sit by a fireplace to keep warm. Hot smoke rises out of the fireplace. You can sit far away from the fire and still feel it warm your skin.

Making Heat Without Using Fuel

There are other ways to make heat. One way people have found to heat homes and buildings is to use the sun. The sun shines through sheets of glass onto a black surface. The sun's light is absorbed and changes to heat. Heat is stored and is used when it is needed. Unlike fuels, we cannot use up the sun's rays. This kind of heating is more useful in places that have many sunny days.

Keeping warm with friction

You may have noticed that you can warm your hands by rubbing them together.

Rubbing your hands together causes **friction.** Friction is a force. Friction changes some of the rubbing motion into heat. Rubbing your hands together makes your hands feel warmer.

CHECKPOINT

1. How do people use heat?
2. How can people make heat without using fuels?

 Where does heat come from?

ACTIVITY

Using the Sun's Heat

Find Out
Do this activity to learn how sunlight makes objects feel warm.

Process Skills
Observing
Communicating
Inferring

WHAT YOU NEED

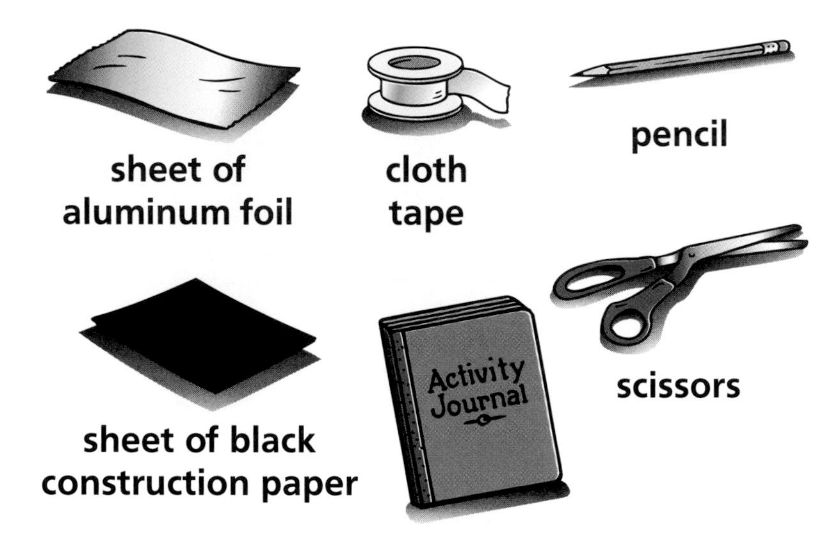

Activity Journal

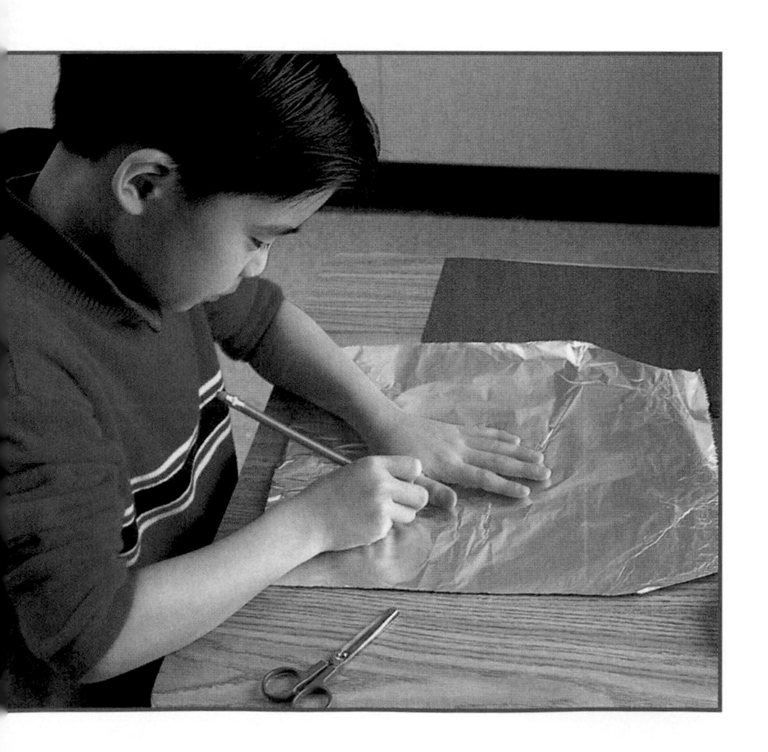

WHAT TO DO

1. Fold the aluminum foil in half, shiny side out. Fold the constuction paper in half.
2. Trace around one hand on the aluminum foil to make a mitten. Trace around the other hand on the constuction paper to make another mitten.

3. Cut out the mittens, cutting through both layers of the foil and paper.

   Safety! **Be careful with scissors.**

4. Keeping the layers together, tape the sides closed. Leave the ends open.

5. Stand outside for five minutes on a sunny day. Keep your eyes closed. Put the mittens on and hold your hands in the sunlight.

   Safety! **Keep your eyes closed.**

6. **Observe** what happens and **record** how you feel.

WHAT HAPPENED

1. Which of your hands felt warmer?
2. Why is this so?

WHAT IF

What would happen if you used plastic wrap instead of aluminum foil?

Effects of Heat

Let's Find Out
- How heat moves
- How heat changes matter
- How heat changes air

Words to Know
conductors
insulators
matter
state
water vapor
evaporation
temperature
thermometers

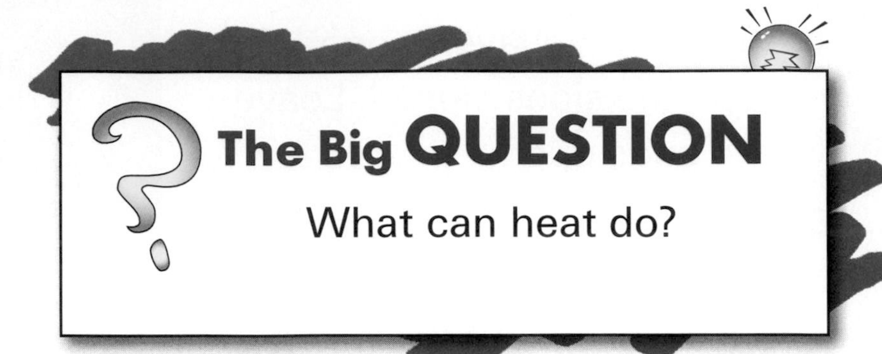

Moving Heat

Heat can move from one object or place to another. Heat moves from warm objects to cooler ones. You can warm your hands by holding a cup of warm soup. Heat moves from the soup through the cup to your hands. You can feel warm air rising above the cup.

Heat moves through some materials more easily than others. Heat moves easily through **conductors.** Most metals are good conductors. Metal pots are used for cooking. Heat from the stove quickly moves through the metal. The heat warms the food.

Other materials are not good conductors. But they may be good insulators. **Insulators** help keep heat from passing through. Most plastics are good insulators. So are clothes you wear, like sweaters and coats. You wear these clothes to keep warm when it is cold outside.

Conductor

Insulators

Changing Matter

Adding or taking away heat can change matter. **Matter** is something that takes up space. Matter can change from one **state,** or form, to another.

An ice cube is solid water. Solid is one state of matter. Heat can melt an ice cube. The ice cube changes into liquid water. Liquid is another state of matter. When heat is taken away, the water can change back. Liquid water turns into solid water.

Heat can make liquids boil. Water boils when it is heated. When the water boils, it turns into a gas. This gas is called **water vapor.** Solid, liquid and gas are three states of matter.

Heat from the sun causes liquid water to turn into water vapor. Water vapor mixes with the air. This is called **evaporation.**

Sometimes heat causes changes that cannot be changed back.

Bread can change into toast when you heat it. Eggs change when you cook them in a pan. You cannot untoast a piece of toast. You cannot uncook an egg.

Changing Air

Heat can warm air, too. This balloon is filled with air. When heat warms the air in the balloon, the air changes. The air takes up more space.

Heat from the sun warms objects all around you, like rocks, streets, and buildings. These objects then warm the air. Warm air is lighter than cold air. Warm air goes up. Cold air takes its place.

You can tell how hot or cold the air is. **Temperature** is a measure of how hot something is. People use **thermometers** to measure the temperature.

CHECKPOINT

1. How does heat move?
2. How can heat change matter?
3. How can heat change air?

 What can heat do?

ACTIVITY

Observing Heat Transfer

Find Out

Do this activity to find out how temperature changes.

Process Skills

Measuring
Communicating
Predicting
Using Numbers
Interpreting Data

What You Need

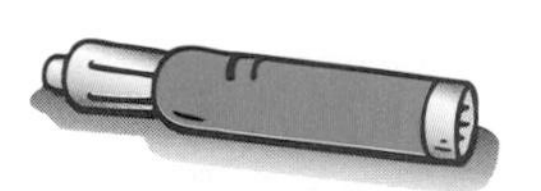

marker

two small cups of water

thermometer

two big cups of water

masking tape

Activity Journal

What to Do

1. Label two small cups "A" and "B." Label two big cups "C" and "D."

2. Fill cups A and B halfway with water at room temperature. Fill cup C halfway with very warm tap water. Fill cup D halfway with very cold water.

Be careful with very warm water.

3. Place the thermometer in cup A. Wait 30 seconds. **Measure** the temperature of the water in this cup and **write** it down. Do the same with the other three cups.
4. Put cup A into cup C. Put cup B into cup D.
5. **Measure** the temperature in A and B every minute for 10 minutes.
6. **Predict** what will happen to the water in A and B after 15 minutes.
7. **Measure** the temperature again after 10 minutes.
8. **Make a graph** to show how the temperature changed.

What Happened

1. How did the temperatures in cup A and cup B change?
2. Why do you think the temperatures changed?

What If

Would an ice cube melt faster if left out in the open air or put into a cup of warm water?

Review

Words to Know

Pick the best word for each sentence.

temperature	**electricity**	**absorb**
light	**water vapor**	**fuel**
friction	**insulators**	**shadow**
state	**heat**	**evaporation**
reflects	**matter**	**conductors**
thermometers		

1. __________ helps us see.

2. A mirror __________ light better than wood.

3. Dark-colored objects __________ more light than light-colored objects.

4. A burning object makes __________.

5. Metal objects are good __________.

6. __________ is a measure of how hot something is.

7. __________ are used to measure temperature.

Using What I Know

1. Does the hat absorb more light or less light than the snow?
2. How will heat change the snow?
3. How can people warm their homes in this kind of weather?

For My **Portfolio**

Write a story about a day when you could not use any fuels. Think about how this would change the way you stay warm, fix meals, and travel.

Unit Review

Telling About What I Learned

1. Objects move in ways that may be predictable. Name two forces that change how an object moves.
2. Sound can be described by its volume and pitch. Give an example of a loud sound, a soft sound, a high sound, and a low sound.
3. Light and heat can travel and be absorbed. Name two sources of light and heat.

Problem Solving

Use the picture to help answer the questions.

1. How will the man and the woman in the picture probably use heat?
2. How are the people in the picture using pushes and pulls?

Something to Do

Make your own band. Make different kinds of musical instruments. Get together with your classmates and play a silly song. Tell why your song is music and not just noise.

UNIT D

Health Science

CHAPTER 1

The Senses

Your senses tell you about your world. You can use your eyes, ears, nose, tongue, and skin to find out about your world. You can find out how objects look, sound, smell, taste, or feel.

The Big IDEA

Your senses and sense organs help you describe the world around you.

CHAPTER SCIENCE INVESTIGATION
Discover the ways you use your senses. Find out what to do in your ***Activity Journal.***

LESSON 1

Your Sense Organs

Let's Find Out

- What your five senses are
- How your sense organs help you find out about your surroundings

Words to Know

senses
organs
eyes
ears
nose
tongue
skin

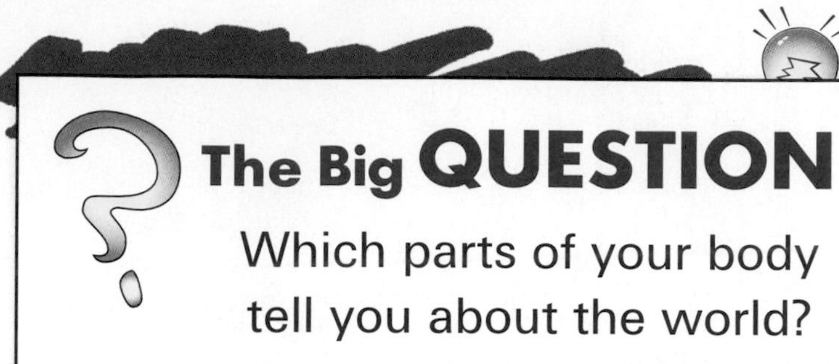

Your Five Senses

Your **senses** tell you most of what you know about the world around you. You have five senses in all. You probably know them. They are sight, hearing, smell, taste, and touch.

Organs are body parts that do special jobs. You have five sense organs. Your **eyes** see. Your **ears** hear. Your **nose** smells. Your **tongue** tastes. Your **skin** feels. What sense organs are the people in the picture using?

Sense Organs and Your Surroundings

Your sense organs tell you things about objects. Your eyes can tell you how tall a maple tree grows and the shape and color of its leaves. Your skin can feel if the bark is rough or smooth, dry or wet. Rough, soft, wet, and dry are kinds of texture. You use size, shape, color, and texture to describe or remember the tree.

Your ears and your eyes are made to sense sound and light. These are forms of energy that travel in waves. For example, when a bird sings, its songs make the air vibrate in waves. Your ear catches the sound waves, and you hear the bird's song. Your eyes catch light waves that help you see the bird.

CHECKPOINT

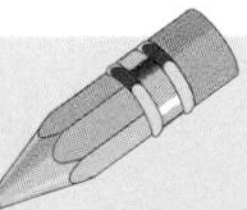

1. What are your five senses?
2. How do you use your sense organs to find out about your surroundings?

Which parts of your body tell you about the world?

ACTIVITY

Using Your Senses

Find Out
Do this activity to see how a sense organ does its job.

Process Skills
Observing
Communicating
Predicting
Inferring

WHAT YOU NEED

two cups of water

a dropper

sugar

salt

Activity Journal

WHAT TO DO

1. **Watch** as your teacher adds salt to one cup of water and sugar to the other cup of water.
2. Place a drop from one cup on your tongue. **Tell** what you taste.

3. Place a drop from the other cup on your tongue. **Tell** what you taste.
4. **Predict** whether or not you can taste with your lips.
5. Place a drop from each cup on your lips. **Tell** what you taste.

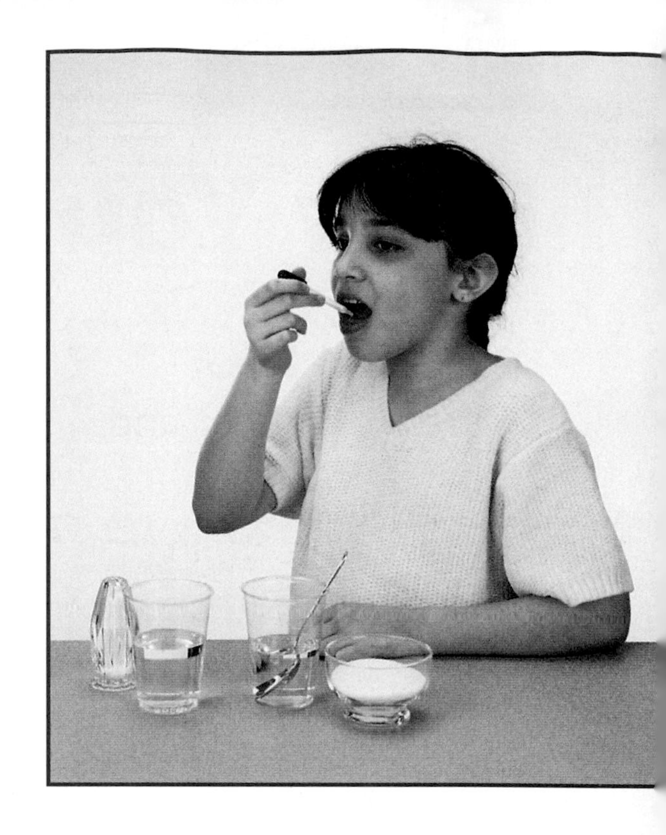

WHAT HAPPENED

1. Did you taste the salt or sugar on your lips? Why do you think this is so?
2. Did you taste the salt or sugar on your tongue? Why do you think this is so?

WHAT IF

What would happen if you put a drop of water on your cheeks? Would you be able to taste the water?

How Your Senses Work

Let's Find Out
- How you hear and taste things
- How your brain gets messages sent by your sense organs

Words to Know
eardrum
vibrate
taste buds
brain
nerves

Hearing and Tasting

Sense organs have many parts you cannot see. One important part of the ear is the **eardrum.** The eardrum is a thin piece of skin inside your ear. Your ear is shaped like a tunnel. The outer part of the ear collects sounds.

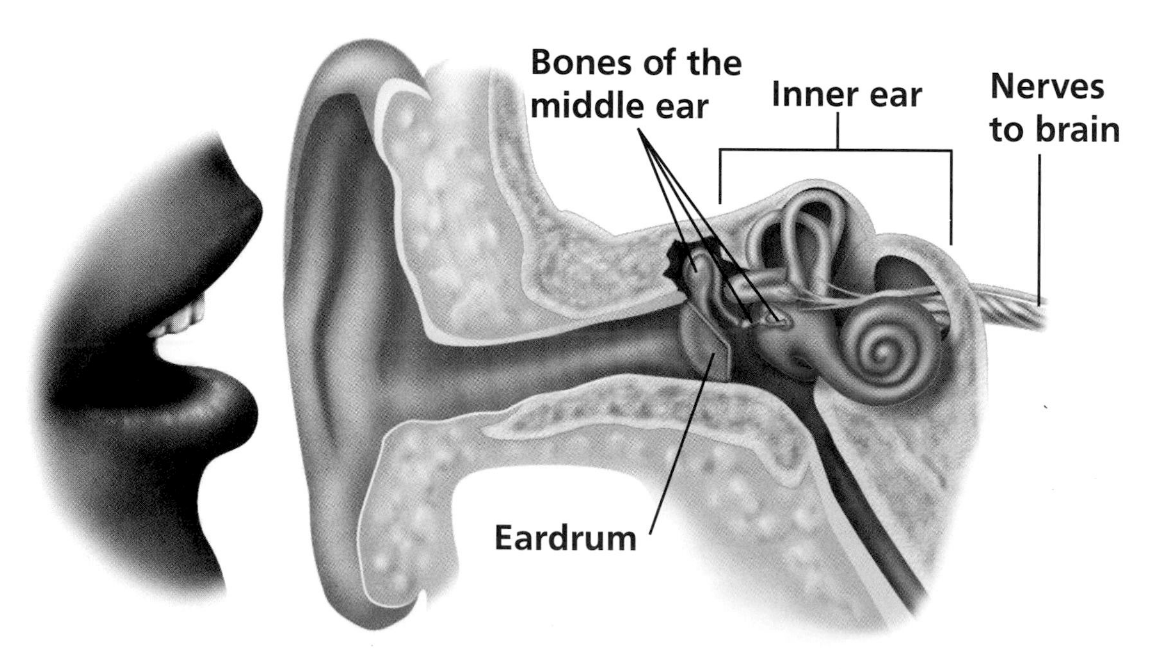

The sounds pass through the tunnel to the eardrum. Once there, the sounds make the eardrum **vibrate,** or move back and forth.

Your tongue is covered with little bumps that contain **taste buds.** Taste buds recognize bitter, sour, salty, or sweet tastes. Everything you taste is some blend of those four tastes.

Apples and oranges taste sweet. Popcorn and pretzels taste salty. Lemons taste sour. Orange rinds taste bitter.

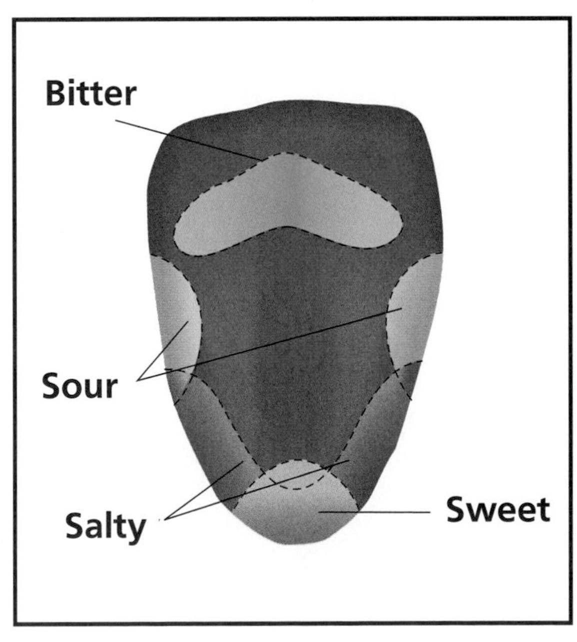

Tongue

Sense Organs and the Brain

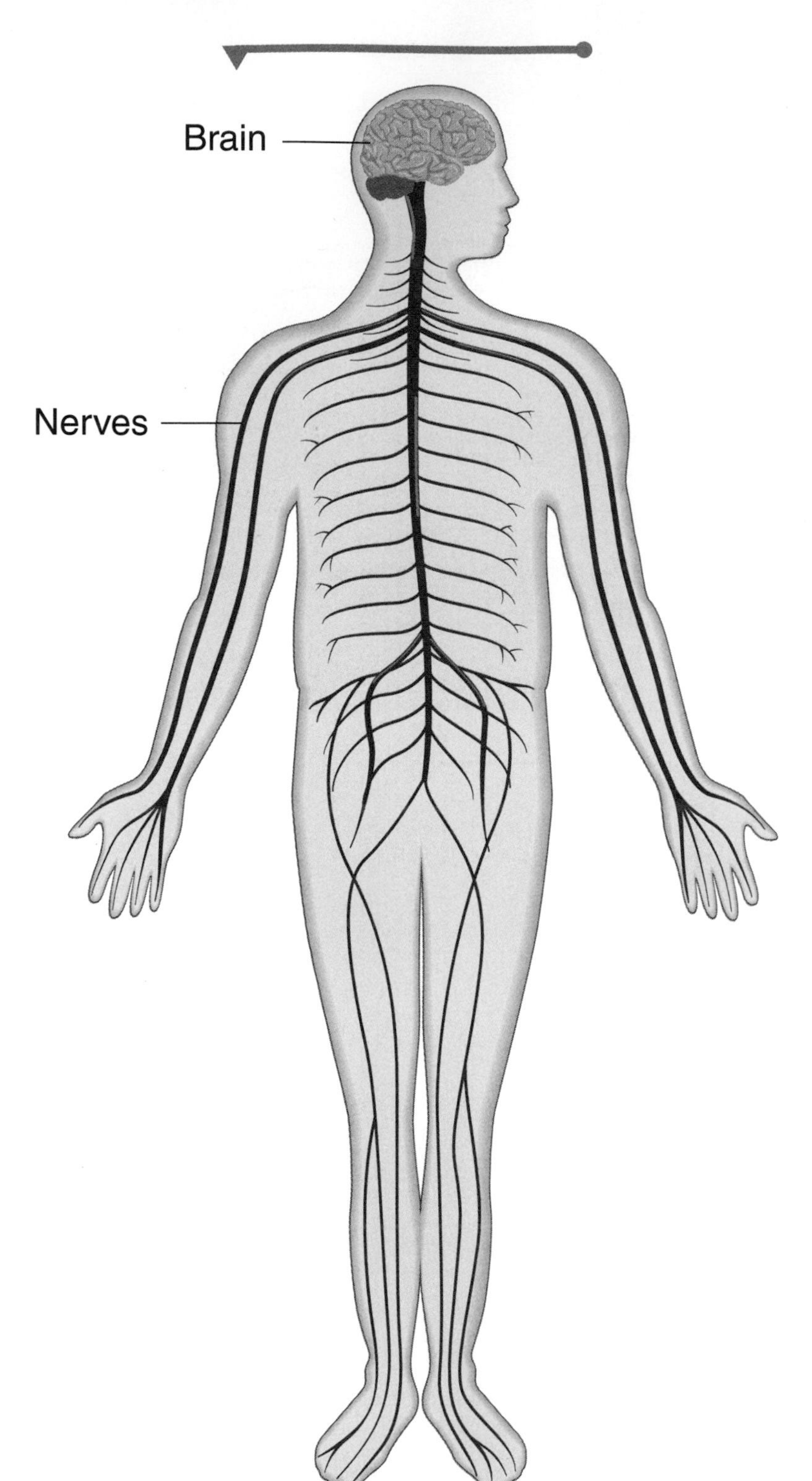

Sense organs do not work alone. They take in information. Then they send it to the **brain** along special pathways, called **nerves.** The brain makes sense of the information. Eardrum vibrations are changed to messages for the brain. Nerves carry them there. Nerves in the taste buds send their own messages. All your sense organs work with your brain in this way.

Smell and taste can work together. Your sense of smell is much more sensitive than your sense of taste. Your nose takes in smells with air. The smells reach nerves high up in your nose. Those nerves signal the brain.

CHECKPOINT

1. How do your ears and taste buds work?
2. How does your brain get messages from your sense organs?

 How do your senses work?

ACTIVITY

Putting Your Senses to Work

Find Out

Do this activity to see how your brain and senses work together with your muscles.

Process Skills

Observing
Communicating
Inferring

WHAT YOU NEED

big hardcover book

coin

clay

Activity Journal

WHAT TO DO

1. With a partner, make five small balls out of clay.
2. Stick the balls on one side of the book to make a simple obstacle course.
3. Put the coin at the start of the course.

4. Hold the book with two hands and move the coin through the obstacle course by tilting the book in different ways. **Observe** what happens.
5. Now close your eyes and have your partner change the course.
6. With your eyes closed, try to move the coin through the new course by **listening** to your partner's instructions.
7. Switch roles with your partner and repeat Step 6.

What Happened

1. What senses were you using when you moved the coin through the course in Step 4? In Step 6?
2. What was the hardest part of doing the activity with your eyes closed? Why?

What If

How could you do this activity if you wore a blindfold and ear plugs?

Senses and Safety

Let's Find Out
- How your senses warn you
- How you use what your senses tell you

Words to Know

warning
danger
protect
safe

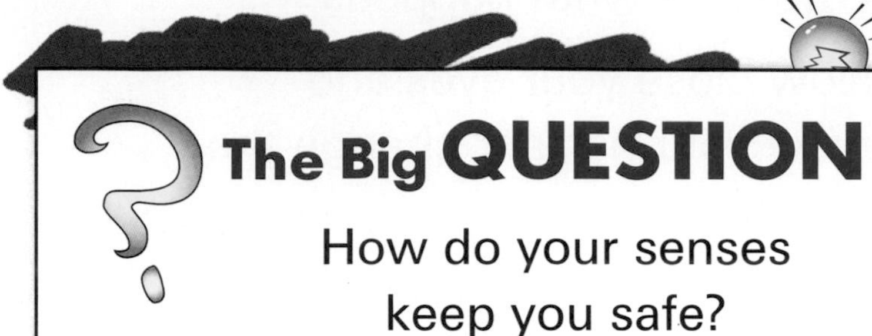

How Your Senses Warn You

Not everything you sense is pleasant. Sometimes your senses give you a **warning.** They alert you to some **danger** around you.

Your senses pick up warnings every day. You see a traffic light turn red. It warns you not to cross the street. You hear a car horn honk or a truck beep. These sounds warn you that a car is coming or a truck is backing up.

Using What Your Senses Tell You

After a fire drill, you wait outside until you hear that it is safe to enter the building.

When your senses warn you, you **protect** yourself. You try to avoid injury and keep yourself **safe** from harm.

If your nose smells smoke, there may be a fire. You walk to an exit as you learned in a fire drill. You leave the building.

When you go swimming, you look around to make sure the area is safe. Before you dive into the water, you look to make sure that there are no obstacles in the water and that the water is deep enough. You decide whether or not you should dive. Water that is too shallow is not safe for diving.

Your senses help protect you in your own home. If you sense that a mug of hot chocolate is too hot, you wait a few minutes before you drink it. You protect your tongue and mouth from a bad burn.

CHECKPOINT

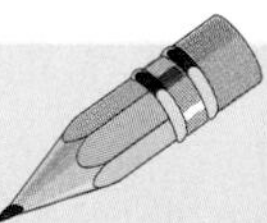

1. How do your senses warn you about danger?
2. How do you use what your senses tell you?

How do your senses keep you safe?

ACTIVITY

Listening for Warnings

Find Out
Do this activity to show how you use your hearing to avoid danger.

Process Skills
Communicating
Observing

WHAT YOU NEED

objects that make sound

Activity Journal

WHAT TO DO

1. Work with a partner.
2. Turn your back to your partner.
3. Have your partner make sounds with four of the objects, one at a time.

4. **Tell** what object makes the sound.
5. **Observe** how you react to each sound. Are you startled? Curious?
6. Switch roles with your partner.
7. Repeat the activity. Use different objects this time.

What Happened

1. Which objects sounded like warnings or alerts? Which did not? Why?
2. How did you feel when you heard the warning sounds?

What If

What other warning sounds can you name?

What would you do if you heard them?

Review

What I Know

Choose the best answer for each sentence.

brain	**organs**	**skin**
warning	**danger**	**eyes**
senses	**eardrum**	**vibrate**
nose	**tongue**	**safe**
ears	**taste buds**	**protect**
nerves		

1. Your __________ is the sense organ that can smell things.

2. Your __________ can feel textures and sense heat and cold.

3. Sound waves make your __________ vibrate.

4. The __________ on your tongue can tell if food is bitter, sour, sweet, or salty.

5. Nerves carry information from your sense organs to your __________.

6. A siren warns you of __________.

7. If you see a hole in the street, you walk around it. That is one example of how your senses __________ you.

Using What I Know

1. Which senses tell the children in the picture about the weather?
2. Which sense organs could warn them that cars and buses are nearby?
3. How can they use the warnings from their senses?

For My Portfolio

Cut out a picture from a magazine. Use three senses to describe the place or object in the picture.

CHAPTER 2

Caring for Your Teeth

You don't think about teeth often. But you use them all the time. You use them to chew food. You use them when you talk to friends. You show them off when you make a big smile.

Your teeth are important to you. With proper care you can help them stay strong and healthy.

The Big IDEA

Teeth help you chew, talk, and look your best.

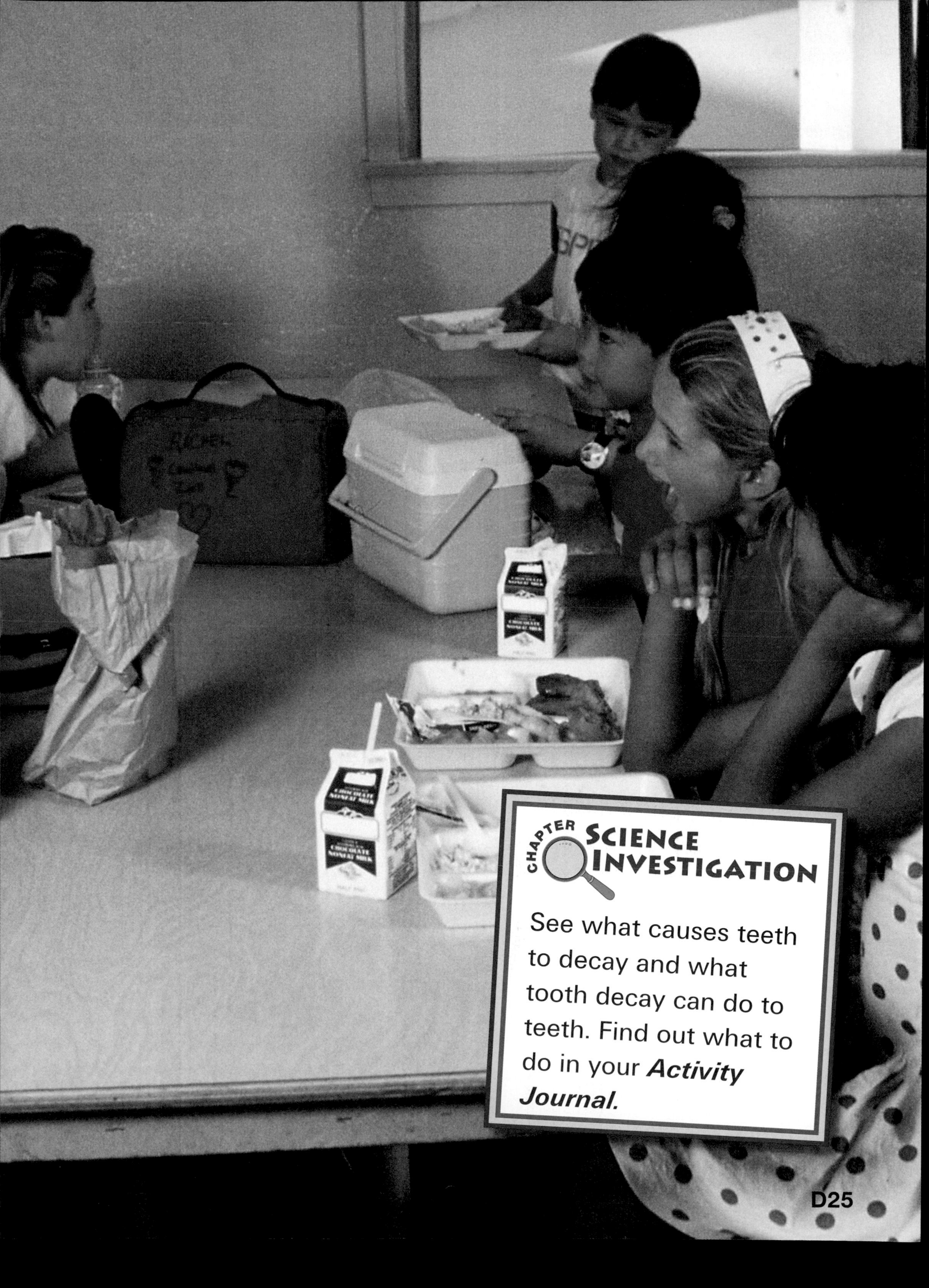

CHAPTER SCIENCE INVESTIGATION

See what causes teeth to decay and what tooth decay can do to teeth. Find out what to do in your ***Activity Journal.***

LESSON 1

Teeth

Let's Find Out

- How you use your teeth
- What kinds of teeth you have

Words to Know

baby teeth
permanent teeth
gum
crown
root

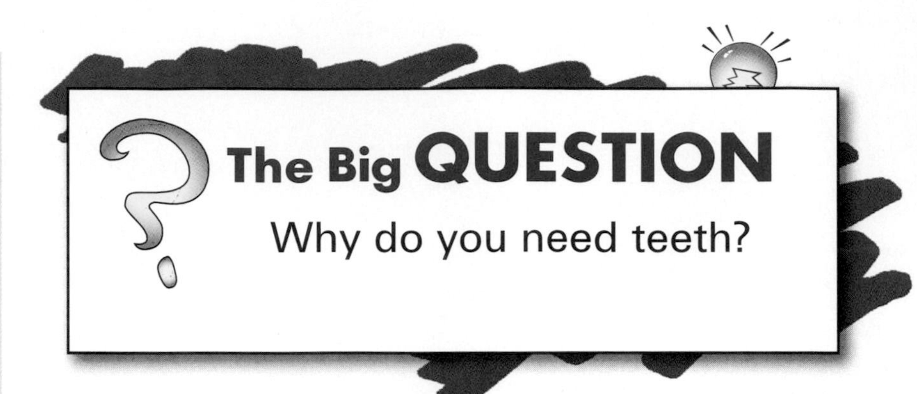

The Purpose of Teeth

What do you use teeth for? If someone asked you that question, you would probably say "For eating." And you would be right. You need teeth to chew solid food. Unless you are eating something soft like yogurt, you use teeth at every meal. You use them for snacking, too.

You also use teeth to speak. Say the word *teeth.* Feel how your tongue presses against your teeth. Without your teeth, it would be hard to say many words.

You need teeth for a nice appearance, too.

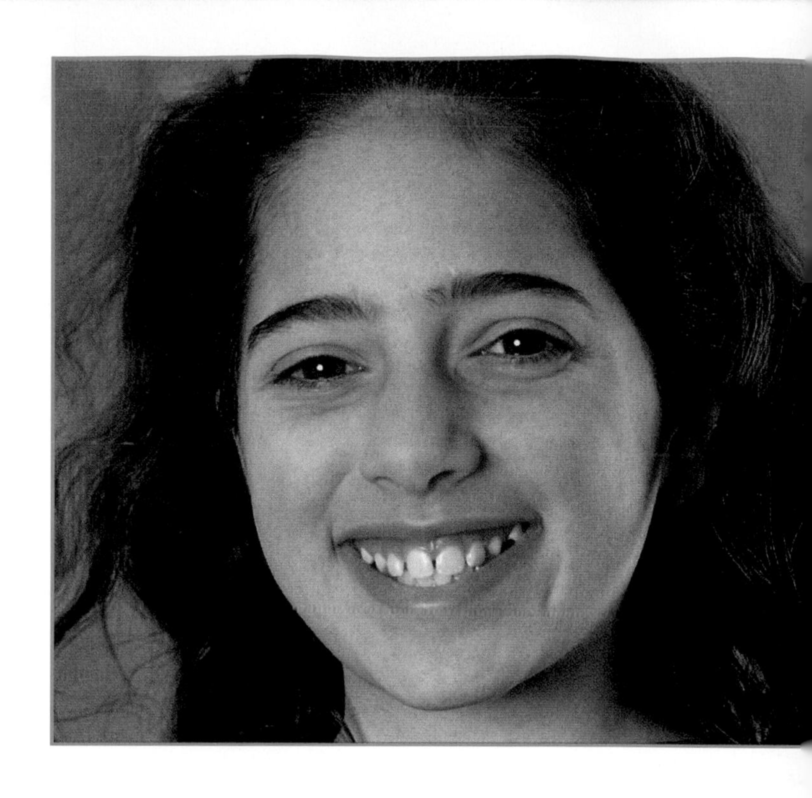

Different Kinds of Teeth

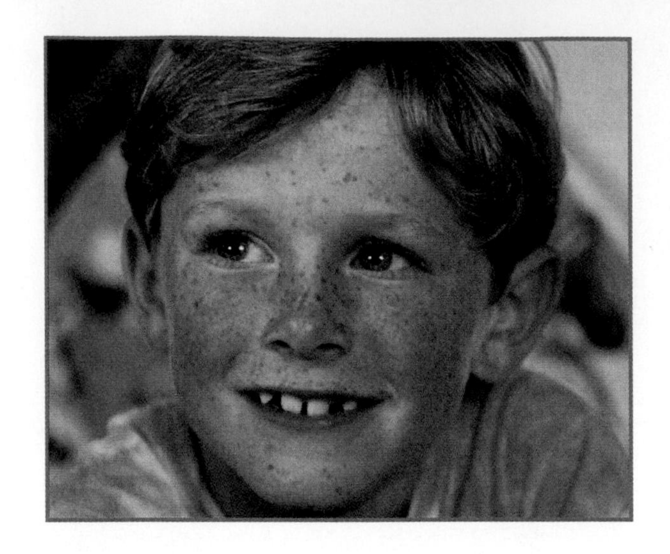

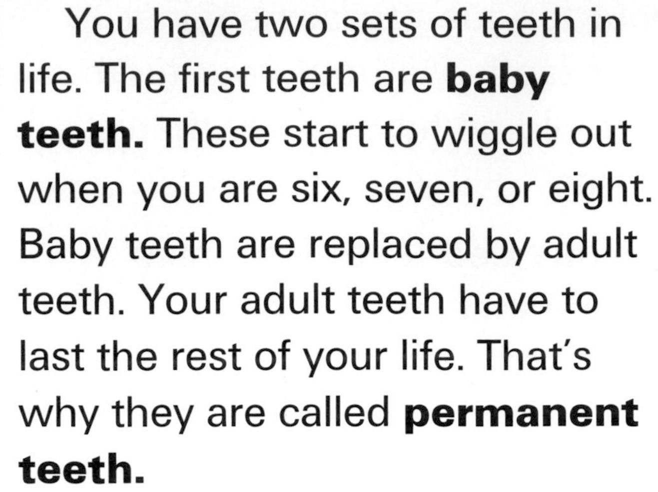

You have two sets of teeth in life. The first teeth are **baby teeth.** These start to wiggle out when you are six, seven, or eight. Baby teeth are replaced by adult teeth. Your adult teeth have to last the rest of your life. That's why they are called **permanent teeth.**

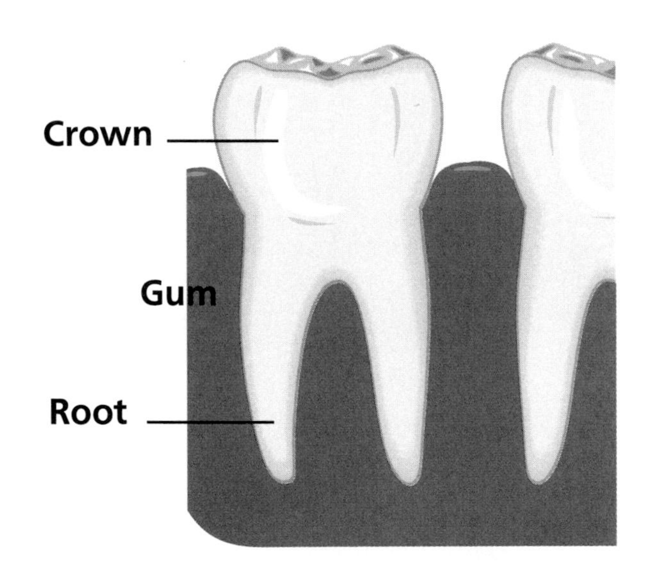

You have soft, pink flesh in your mouth that surrounds the bottom of your teeth. That flesh is called the **gum.** What you see in your mouth is just part of a tooth. The rest is hidden below the gum. The part of the tooth you see is called a **crown.** The hidden part is called a **root.** The root holds the tooth in your mouth.

Permanent teeth start to grow in your gum while you still have your baby teeth. As these new teeth grow, they push the baby teeth out. Permanent teeth are larger than baby teeth. If you take care of them, you will have them for the rest of your life.

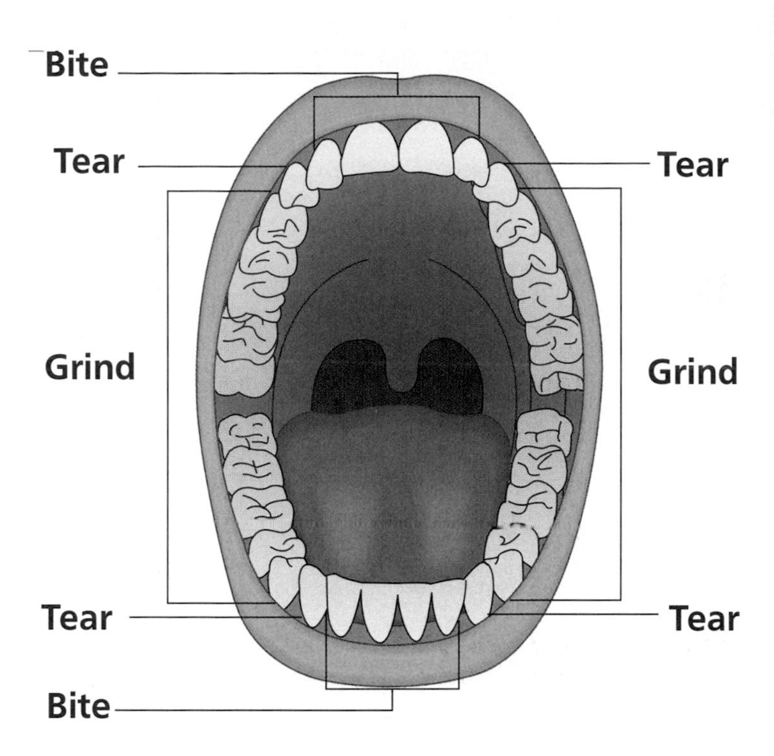

You have different types of teeth. They do different jobs in chewing. Your front teeth bite into food. Your pointed side teeth tear the food. The short, thick teeth at the back of your mouth grind food. Your back teeth are the strongest. They do most of the chewing.

CHECKPOINT

1. How do you use teeth to eat?
2. How do your teeth do different jobs?

 Why do you need teeth?

ACTIVITY

Using Teeth to Speak

Find Out
Do this activity to see how your teeth help you speak clearly.

Process Skills
Communicating
Observing

WHAT YOU NEED

paper

flash cards with letters f, l, v, and th

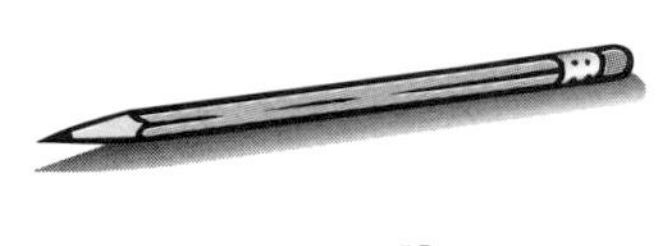

pencil

Activity Journal

WHAT TO DO

1. Work in four groups.
2. Each group takes a flash card.
3. Think up words that use the letter(s) on your flash card. Help your group **write** five words.

4. **Read** your words aloud to the class.
5. Cover your teeth with your lips. **Observe** how the sound changes.
6. Now try to **say** the same words.

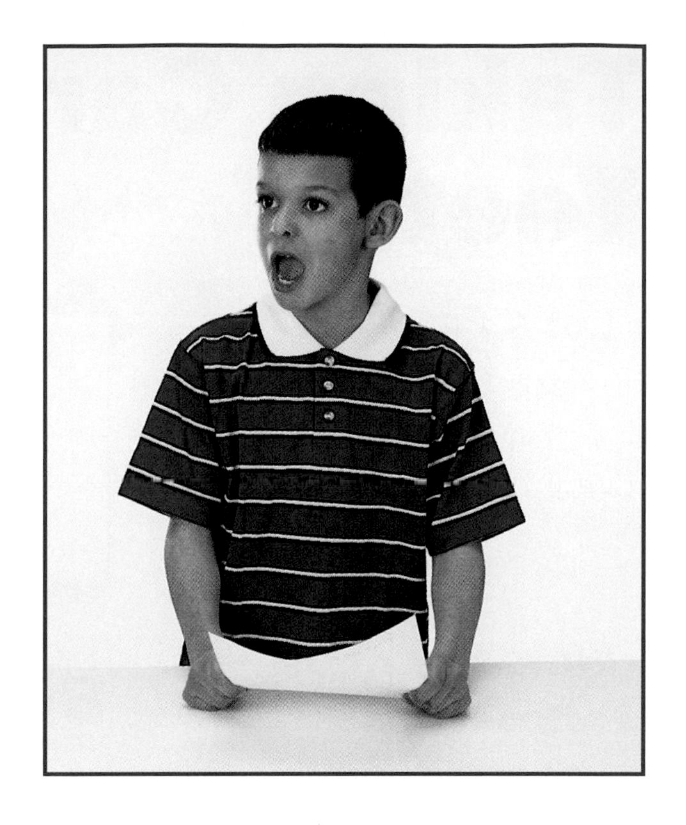

What Happened

1. When could you speak your words clearly?
2. When couldn't you? Why not?

What If

Name some words made up of sounds that don't use teeth. Hint: Babies say them!

Taking Care of Your Teeth

Let's Find Out
- How to brush and floss your teeth
- What causes tooth decay

Words to Know
floss
tooth decay
cavity

Brushing and Flossing

Caring for teeth is important. You have to do your part. Brush your teeth at least twice a day. Try to brush your teeth after every meal. Here's how you should brush your teeth.

Brush down on your top teeth, and up on your lower ones, so the toothbrush bristles don't push back your gums. Then, move the brush in small circles across the teeth.

Next, brush the inside of your teeth and the chewing surfaces. Don't forget your back teeth! Make sure you get all the food out of the grooves.

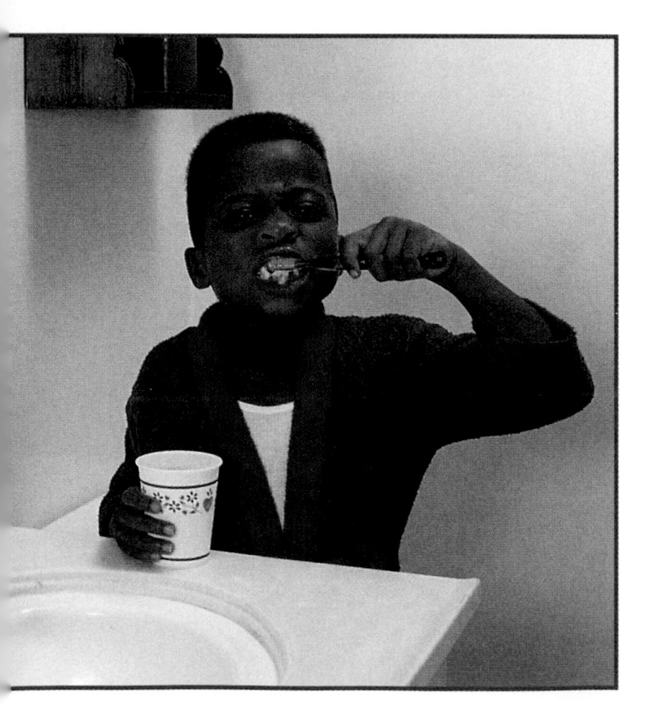

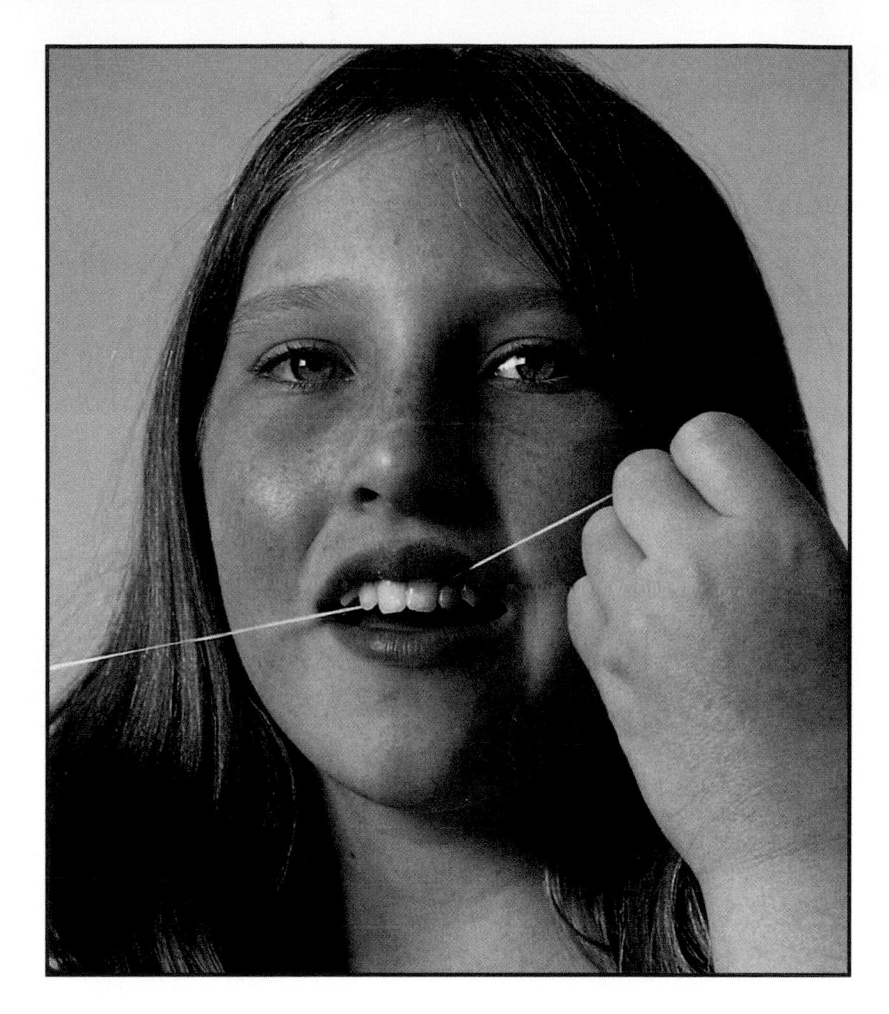

You should **floss** your teeth once a day. Floss is a special string that cleans between your teeth. It gets out food that you can't reach with a toothbrush.

To floss, wrap the thread around your fingers. Then ease it between your teeth. Move the floss down to the gum with a gentle back and forth movement. Repeat between all your teeth. Rinse with water after flossing and brushing your teeth.

Tooth Decay

Adults may tell you not to eat too many sweets. There's a reason for it. Sugar may taste good, but it's not good for your teeth.

Sugar plays a role in **tooth decay.** Decay means "to rot." Tooth decay happens when you don't take care of your teeth properly. This is what happens.

Food left in your mouth combines with germs. A sticky coating forms. This coating clings to your teeth.

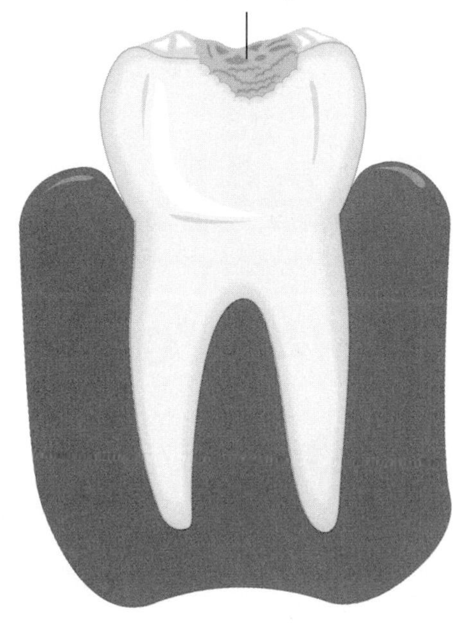

The germs in the coating cause your teeth to begin to decay. In time, the decay can make a hole in your teeth. This hole is called a **cavity.**

Flossing and brushing your teeth help take off the sticky coating. That's why they are important. You should rinse your mouth with water after eating sugary foods if you cannot brush your teeth.

CHECKPOINT

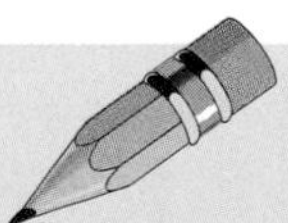

1. How should you brush and floss your teeth?
2. What causes tooth decay?

 How should you care for your teeth?

ACTIVITY

Flossing Teeth

Find Out
Do this activity to see why flossing is important.

Process Skills
Predicting
Observing

WHAT YOU NEED

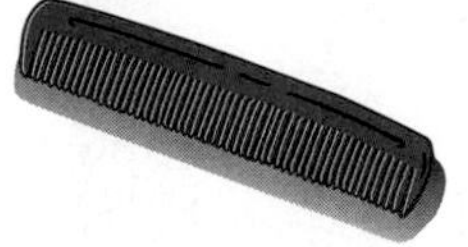
comb

craft stick

yarn

cream cheese

paper towels

Activity Journal

WHAT TO DO

1. Spread the cream cheese on the comb with the craft stick.
2. Wipe the comb with the paper towel. **Predict** how much of the cream cheese will be wiped off.

3. Now use the yarn to clean between the teeth of the comb. **Observe** what happens.

What Happened

1. Did the paper towel clean all the cream cheese from the comb?
2. How did the yarn help clean the comb?

What If

How is this activity like brushing and flossing your teeth?

Going to the Dentist

Let's Find Out

- How the dentist helps you
- What X rays and fillings are

Words to Know

fluoride
tooth X ray
filling

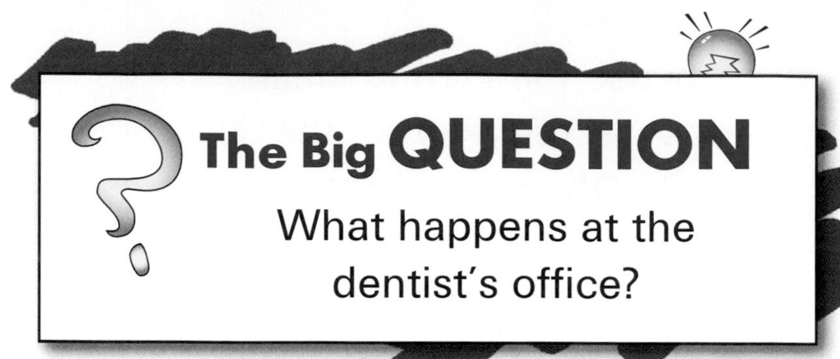

The Dentist

A dentist helps keep your teeth healthy. You should go to the dentist twice a year. The dentist gives you care you can't give yourself. When you visit the dentist, here are some things you can expect.

The dentist will examine your mouth. The dentist will also check for cavities. The dentist will probably tell you how to take better care of your teeth.

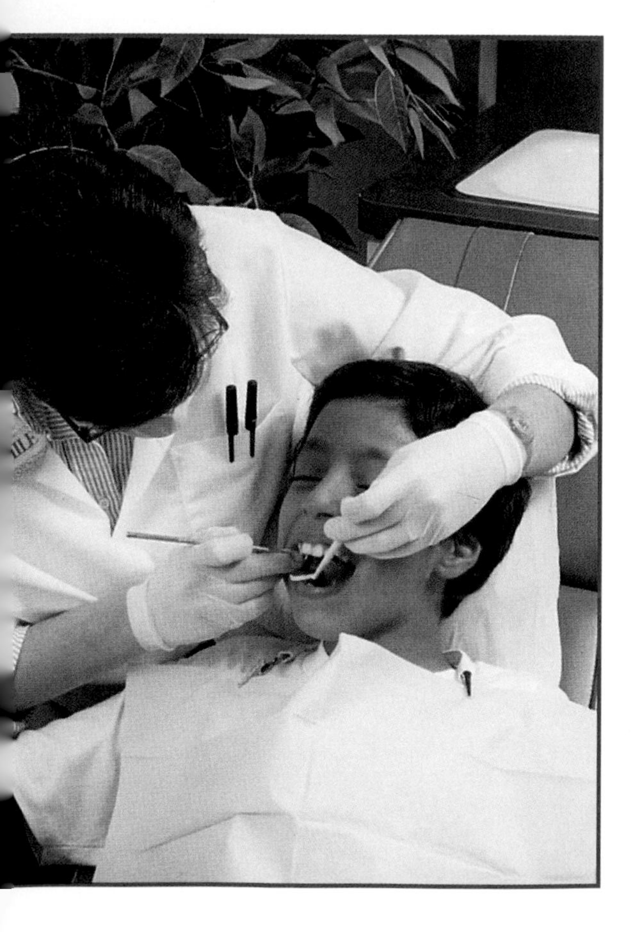

A dental hygienist will help the dentist take care of your teeth. The hygienist will clean and polish your teeth. The hygienist will use a machine at the dentist's office that cleans the sticky coating off your teeth. The hygienist has tools to clean your teeth better than you can clean them.

The dental hygienist may put **fluoride** (flōr′ īd) on your teeth. Fluoride is a substance that helps make teeth strong and less likely to get cavities. Fluoride is found in most drinking water and toothpastes.

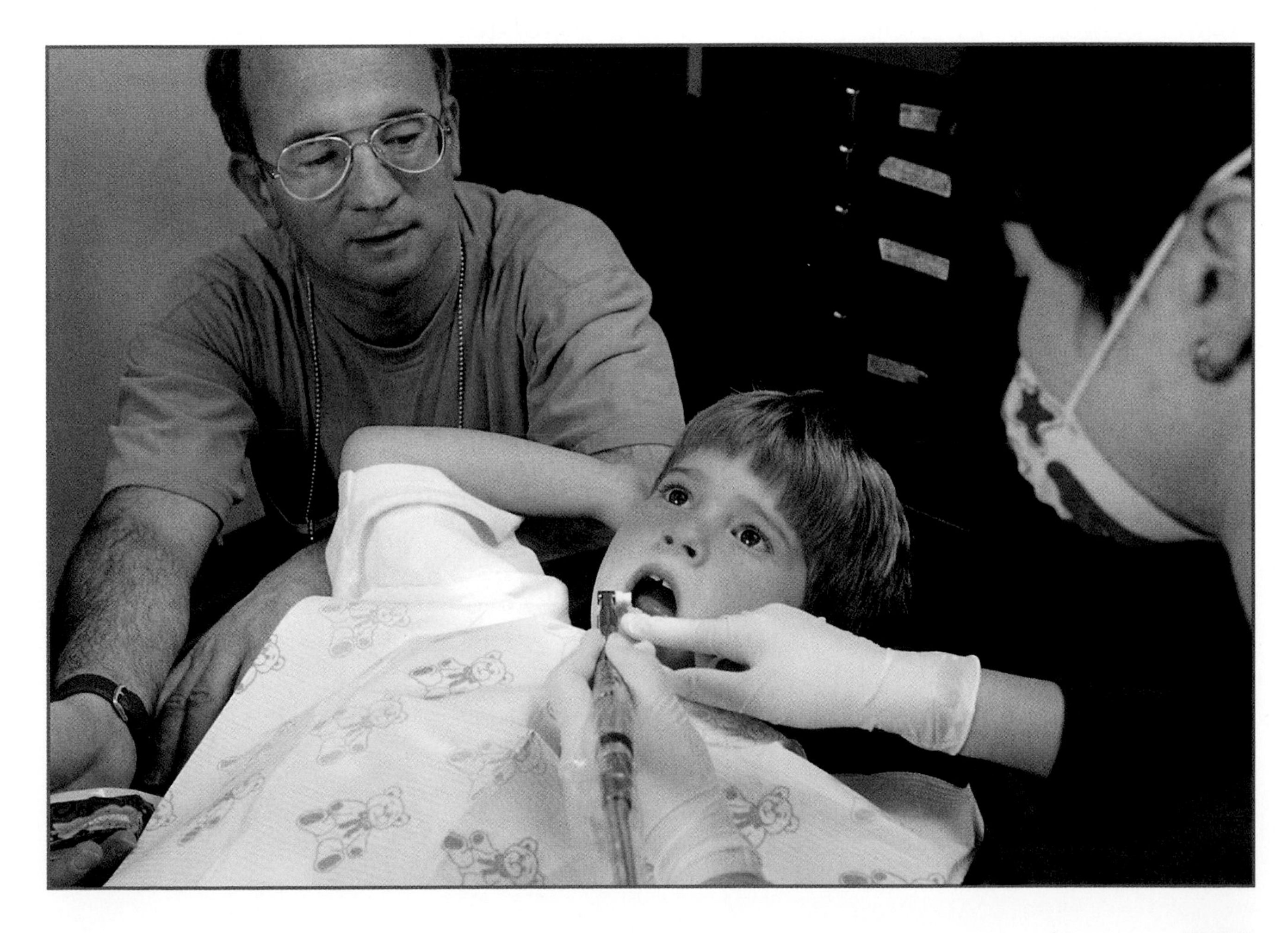

X Rays and Fillings

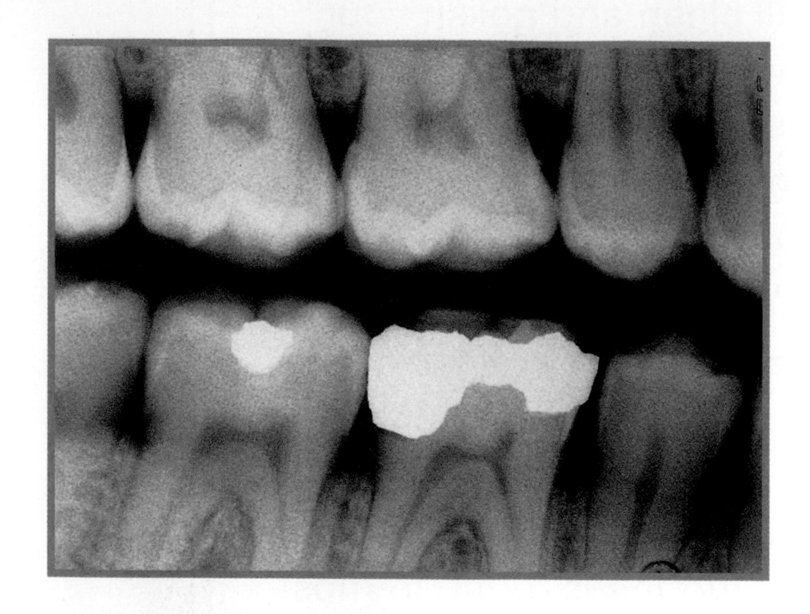

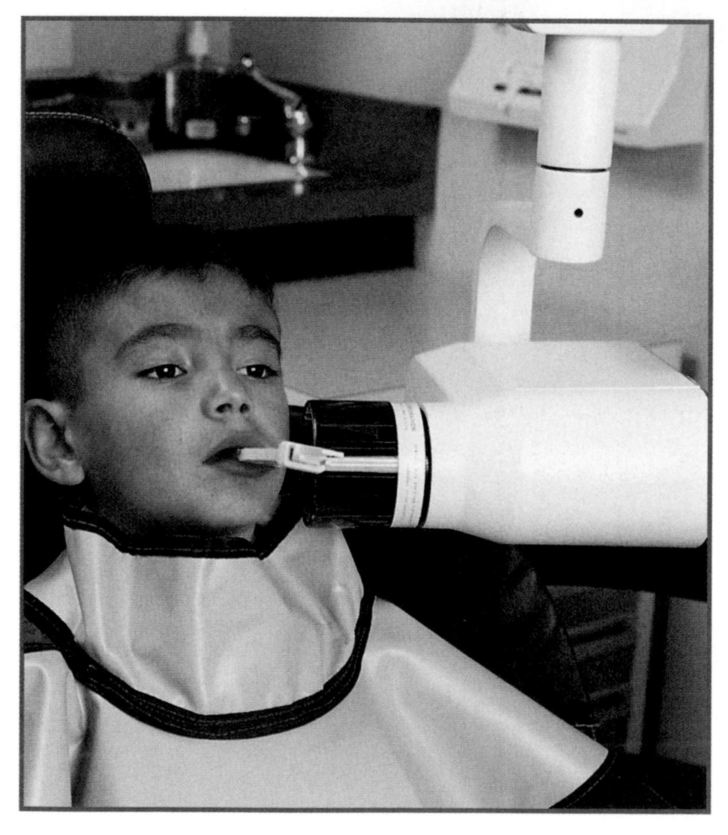

The dental hygienist or a dental assistant may take pictures of your teeth with a special machine. A picture of your teeth is called a **tooth X ray.** X rays are different from ordinary pictures. They show the insides of teeth. X rays help the dentist to see tiny holes in your teeth that cannot be seen easily.

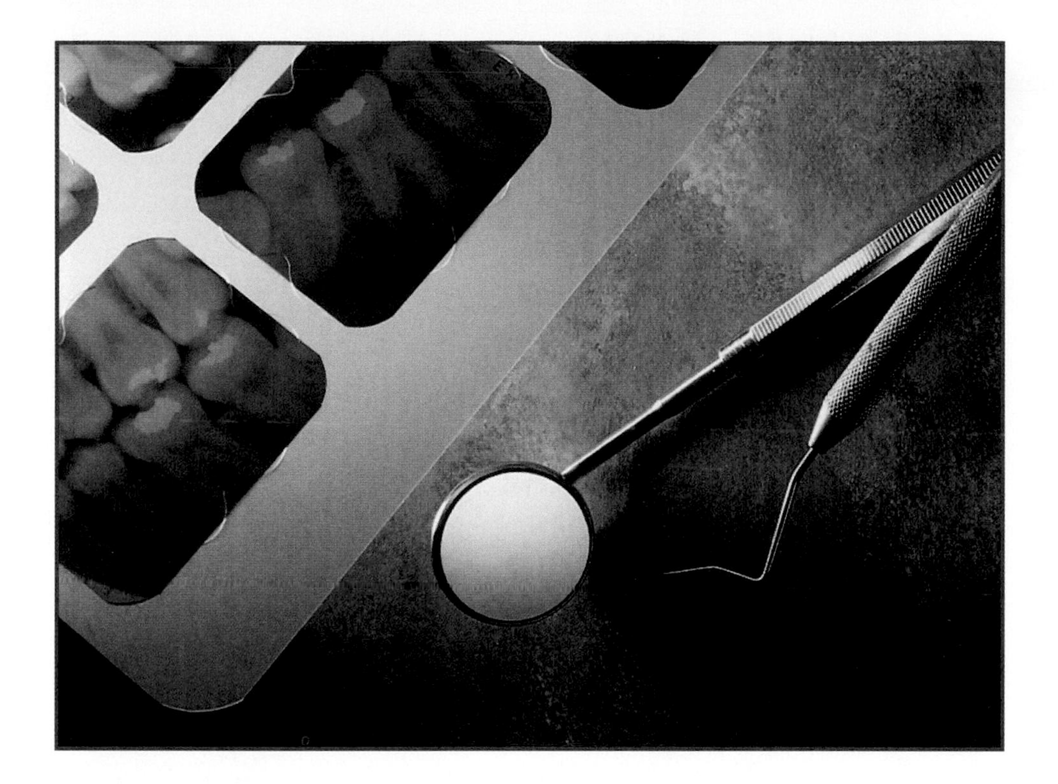

What happens if you have a cavity? First, the dentist will clean out the cavity. Cleaning stops the decay from working deeper into your tooth.

Then, the dentist will put in a **filling.** A filling fills the hole that is made by the cavity.

CHECKPOINT

1. How does the dentist help you?

2. How does an X ray help the dentist?

 What happens at the dentist's office?

ACTIVITY

Fixing Teeth with Fillings

Find Out
Do this activity to learn how a dentist fixes a cavity.

Process Skills
Constructing Models
Predicting

WHAT YOU NEED

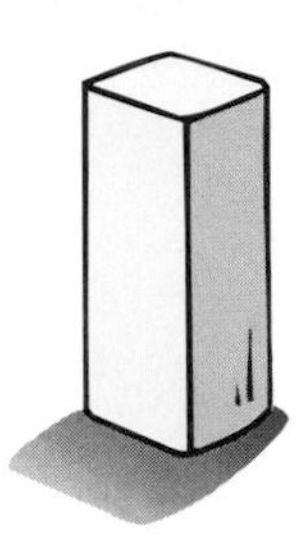

yellow modeling clay

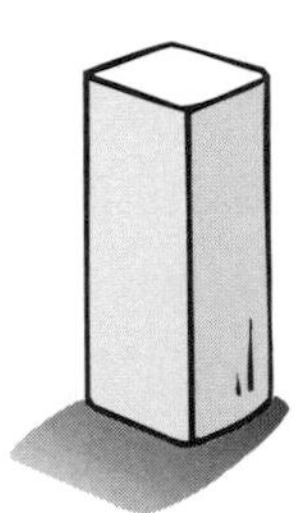

white modeling clay

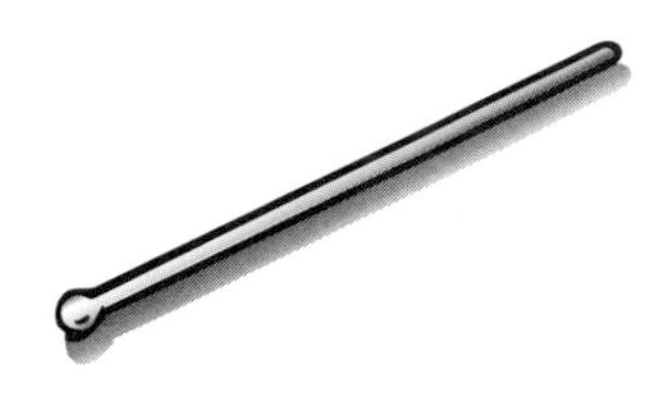

plastic stirrer

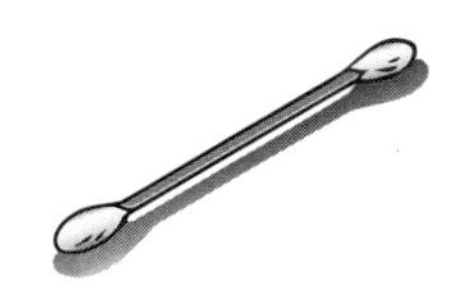

cotton swabs

Activity Journal

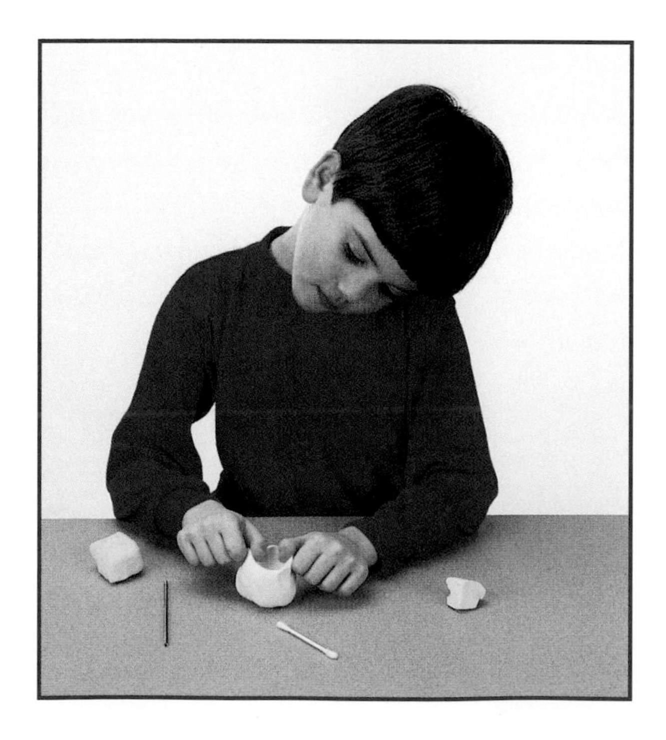

WHAT TO DO

1. **Make** a cupcake-size tooth with the white clay.
2. **Make** a hole in the tooth with the plastic stirrer.

3. Clean the hole with the cotton swab. **Predict** how much yellow clay you will need to fill the hole.
4. Fill the hole with yellow clay.

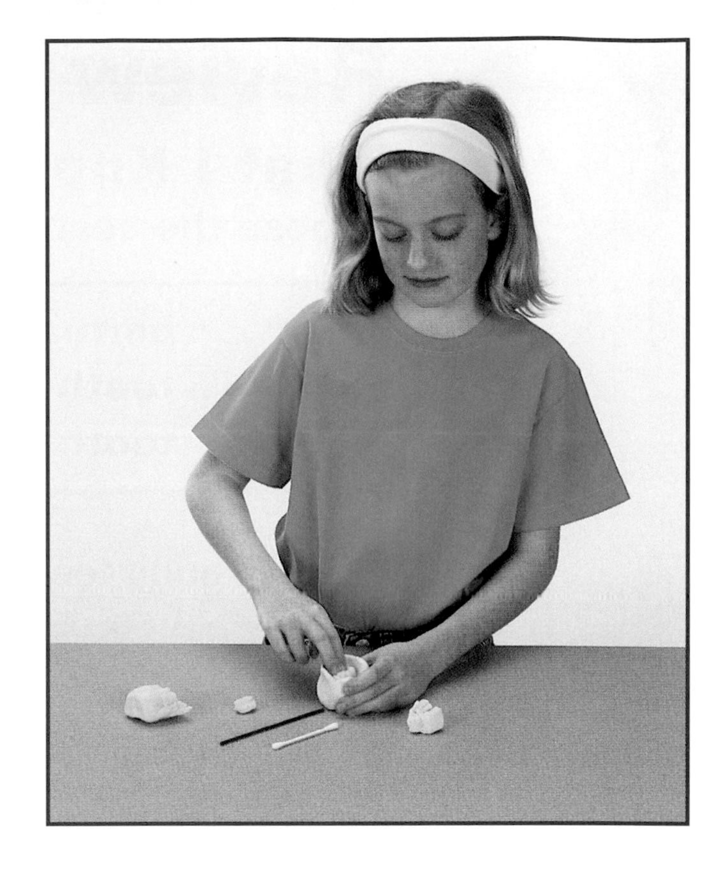

What Happened

1. Why was the yellow clay put in the hole?
2. How was this activity like a dentist fixing a cavity?

What If

What would happen if the hole in the tooth got bigger and bigger?

Review

What I Know

Choose the best answer for each sentence.

root	**permanent teeth**	**filling**	**crown**
cavity	**tooth X ray**	**gum**	**fluoride**
floss	**tooth decay**	**baby teeth**	

1. Your adult teeth are called __________ because they have to last the rest of your life.

2. The soft, pink flesh in your mouth that surrounds the base of the teeth is called the __________.

3. The part of the tooth that is hidden below the gum is called the __________.

4. You should __________ once a day to clean between your teeth.

5. __________ happens when you don't take care of your teeth properly.

6. A hole in a tooth is called a __________.

7. A __________ helps a dentist see holes in teeth.

Using What I Know

1. Why do you think the boy in the picture has missing teeth?

2. What will appear next in the spaces between his teeth?

3. How can he take care of his teeth?

For My *Portfolio*

Draw a picture of a tooth with the gum, crown, and root. Label the gum, crown, and root on your picture.

You are what you eat. That means the food you eat has an effect on your body. Food helps you work, play, and think. Not all food is the same. You need to eat enough of the right foods to stay healthy.

The Big IDEA

A healthful diet helps you grow and stay well.

CHAPTER SCIENCE INVESTIGATION
Keep track of your diet. See how what you eat makes you feel. Find out how in your *Activity Journal.*

LESSON 1

The Food Guide Pyramid

Let's Find Out

- How to eat a balanced diet
- How the Food Guide Pyramid can help you plan a healthful meal
- How many servings you should eat

Words to Know

diet
Food Guide Pyramid
serving
nutrients

The Big QUESTION

How much of each food group should you eat?

A Balanced Diet

Your **diet** is what you eat regularly. People eat many kinds of food. Some people eat a lot of fruits and vegetables, which come from plants. Other people eat a lot of meat and eggs, which come from animals.

A balanced diet is one that gives your body everything it needs. A balanced diet has many different kinds of foods.

Each food belongs to a food group. For good health, you need foods from each of the food groups.

Milk, yogurt, and cheese group

Vegetable group

Bread, cereal, rice, and pasta group

Fruit group

Meat, poultry, fish, eggs, dry beans, and nuts group

The Food Guide Pyramid

The best way to get what you need is to follow the **Food Guide Pyramid.** It is a guide to healthful eating. The Food Guide Pyramid tells you how many foods from each group you need to eat each day.

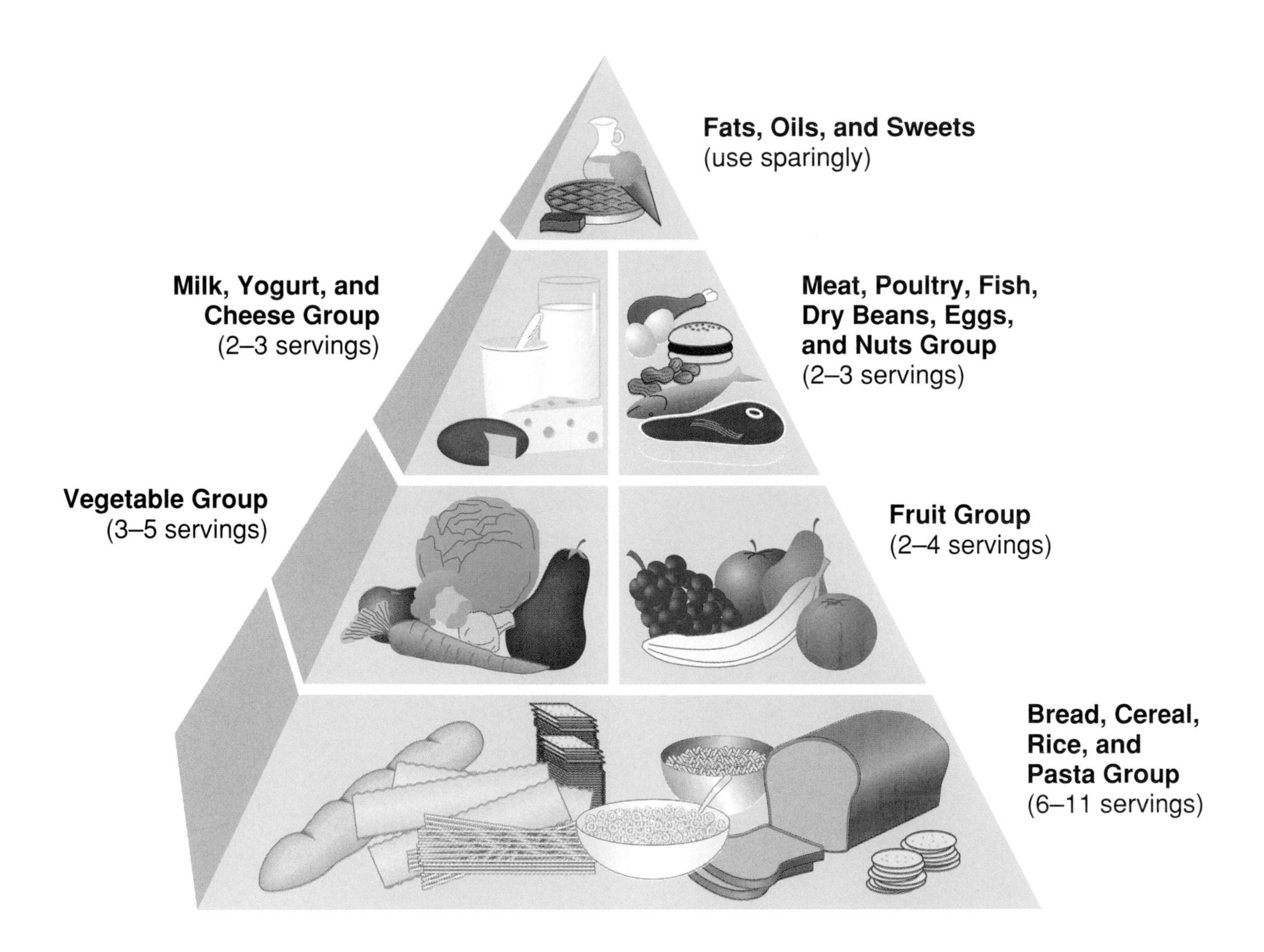

The size of each section in the Food Guide Pyramid shows how much of that kind of food you should eat each day. You can tell that you should not eat many fats, oils, or sweets. Did you eat enough fruits and vegetables today? Look at the guide.

You do not have to eat the same thing every day to have a balanced diet. If you ate chicken yesterday, you might eat fish or eggs today. If you ate pasta last night, you might eat rice tonight. These foods don't taste the same, but they help you in similar ways. A healthful diet doesn't have to be boring. When you eat a lot of different foods, your meals taste great and are fun!

How Many Servings

A **serving** is a suggested amount of food. If you read the back of a food package, you can see how many servings it has. You can also find out the kinds of **nutrients** that the food has. A nutrient is a substance in food that helps your body work and play.

The Food Guide Pyramid shows you how many servings of each kind of food you should eat each day. One slice of bread or a half cup of rice is one grain serving. You should eat at least six grain servings each day. One apple or three-fourths cup of juice is a fruit serving. You should eat at least two fruit servings each day.

Almost all foods can be part of a healthful diet. You should not eat too much of any kind of food. But, you need to eat enough food so that you have enough energy to work and play. The Food Guide Pyramid can help you choose what kinds of food you should eat and how much of each kind you should eat. You can even use it to choose a healthful snack.

CHECKPOINT

1. How can you eat a balanced diet?
2. How can the Food Guide Pyramid help you plan a healthful meal?
3. How many grain servings should you eat each day? How many fruit servings?

How much of each food group should you eat?

ACTIVITY

Making a Balanced Meal

Find Out
Do this activity to learn how to make a balanced meal.

Process Skills
Communicating
Classifying

WHAT YOU NEED

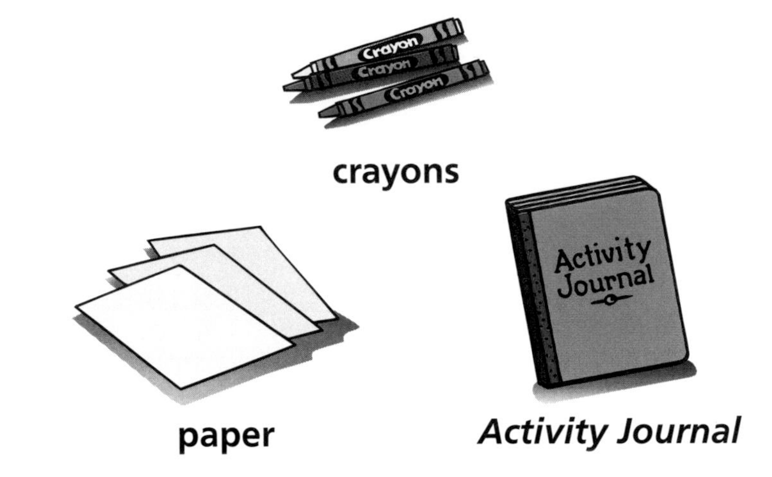

WHAT TO DO

1. Plan three meals that use all of the food groups. Include some new foods you want to try.
2. **Draw** pictures of the foods that you choose.

3. **Arrange them** together in three groups as meals.
4. Do not use the same food twice.

What Happened

1. In which food group is your favorite food?
2. What new foods did you include?

What If

What could you do if you think that your diet is not healthful?

How Your Body Uses Food

Let's Find Out

- How your body gets energy
- What different foods do for your body

Words to Know

energy

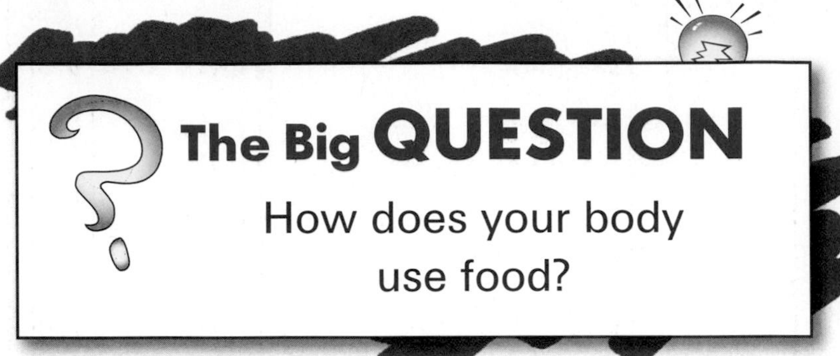

Energy

Energy is your ability to work or play. You are using energy right now. Reading uses energy. Sleeping uses energy. Even breathing and growing use energy. Some activities use more energy than others. Running uses more energy than playing a computer game or reading a book. Eating food is how your body gets the energy it needs.

The food you eat is broken down in your body. You chew your food into smaller pieces. Then, the food goes into your stomach. There it becomes even smaller. Finally, all that is left are the nutrients your body needs. The bloodstream absorbs nutrients. Then, the bloodstream carries the nutrients to different places. The body uses the nutrients for energy and to grow.

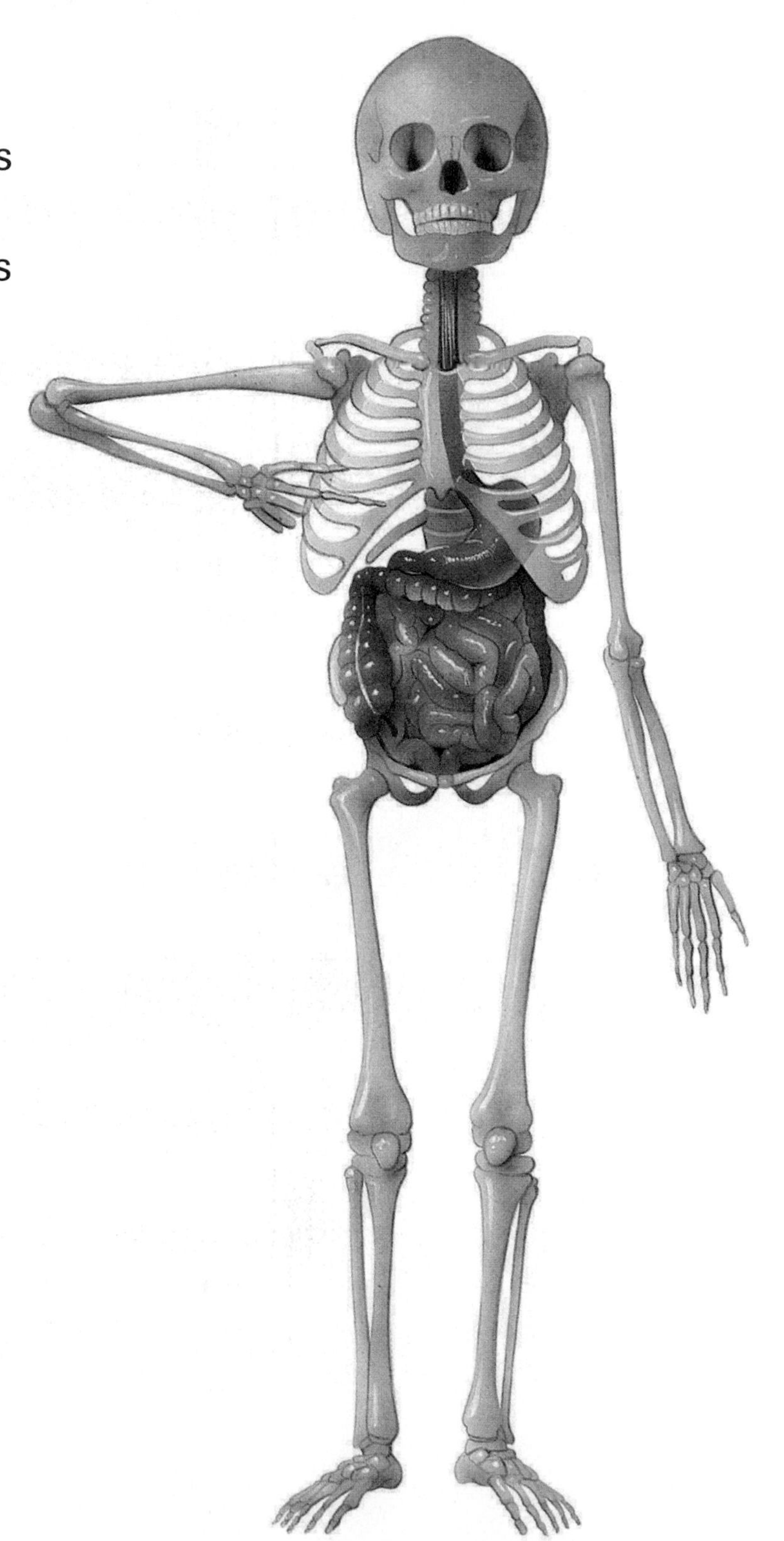

Different Foods

If you want to eat well, you need to know what your body needs. You also need to know what different foods can do for you. Not all foods are the same. Your body uses different kinds of foods in different ways.

Soda and chips might taste good, but they are not the best choices for a balanced diet. They do not have everything your body needs.

Fruit gives the body quick energy. Pasta gives the body energy it can use over a long time. The calcium in milk and yogurt builds strong bones and teeth. The vitamins in fruits and vegetables also help keep your body healthy.

Sugar and candy have no nutrients to give your body. Fruit, yogurt, and peanut butter taste great and have nutrients to help your body grow.

CHECKPOINT

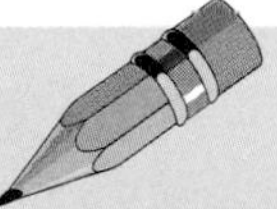

1. How does your body get energy?
2. What kinds of food give your body quick energy? What kinds of food give your body energy that lasts a long time?

 How does your body use food?

ACTIVITY

Learning About Foods

Find Out

Do this activity to find out more about the foods you eat.

Process Skills

- Observing
- Using Numbers
- Inferring
- Interpreting Data

WHAT YOU NEED

healthful-cereal box

cupcake wrapper

sugared-cereal box

soup label

Activity Journal

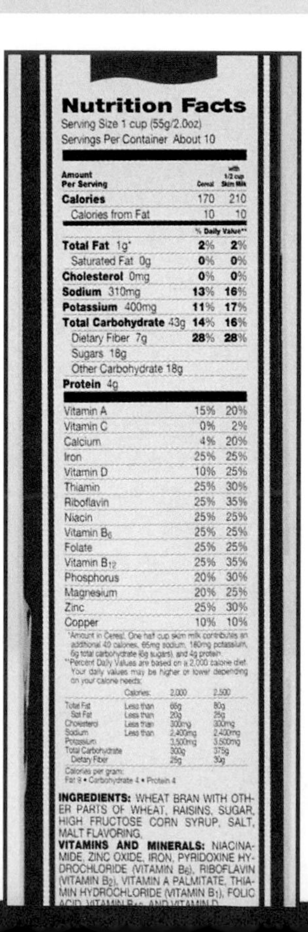

Nutrition Facts

Serving Size 1 cup (55g/2.0oz)
Servings Per Container About 10

Amount Per Serving	Cereal	with 1/2 cup Skim Milk
Calories	170	210
Calories from Fat	10	10
	% Daily Value**	
Total Fat 1g*	**2%**	**2%**
Saturated Fat 0g	**0%**	**0%**
Cholesterol 0mg	**0%**	**0%**
Sodium 310mg	**13%**	**16%**
Potassium 400mg	**11%**	**17%**
Total Carbohydrate 43g	**14%**	**16%**
Dietary Fiber 7g	**28%**	**28%**
Sugars 18g		
Other Carbohydrate 18g		
Protein 4g		
Vitamin A	15%	20%
Vitamin C	0%	2%
Calcium	4%	20%
Iron	25%	25%
Vitamin D	10%	25%
Thiamin	25%	30%
Riboflavin	25%	35%
Niacin	25%	25%
Vitamin B_6	25%	25%
Folate	25%	25%
Vitamin B_{12}	25%	35%
Phosphorus	20%	30%
Magnesium	20%	25%
Zinc	25%	30%
Copper	10%	10%

*Amount in Cereal. One half cup skim milk contributes an additional 40 calories, 65mg sodium, 180mg potassium, 6g total carbohydrate (6g sugars), and 4g protein.
**Percent Daily Values are based on a 2,000 calorie diet. Your daily values may be higher or lower depending on your calorie needs:

	Calories:	2,000	2,500
Total Fat	Less than	65g	80g
Sat Fat	Less than	20g	25g
Cholesterol	Less than	300mg	300mg
Sodium	Less than	2,400mg	2,400mg
Potassium		3,500mg	3,500mg
Total Carbohydrate		300g	375g
Dietary Fiber		25g	30g

Calories per gram:
Fat 9 • Carbohydrate 4 • Protein 4

INGREDIENTS: WHEAT BRAN WITH OTHER PARTS OF WHEAT, RAISINS, SUGAR, HIGH FRUCTOSE CORN SYRUP, SALT, MALT FLAVORING.
VITAMINS AND MINERALS: NIACINAMIDE, ZINC OXIDE, IRON, PYRIDOXINE HYDROCHLORIDE (VITAMIN B_6), RIBOFLAVIN (VITAMIN B_2), VITAMIN A PALMITATE, THIAMIN HYDROCHLORIDE (VITAMIN B_1), FOLIC ACID, VITAMIN B_{12}, AND VITAMIN D.

WHAT TO DO

1. **Look** at the sides of the packages. **Look** for nutrition facts or nutrition information.
2. Read the serving size on each package.

3. **Find the number** of Calories that are in each serving.
4. Read the information about the vitamins and calcium.
5. **Make a graph** that shows how many Calories are in each food.

What Happened

1. Which food has the most Calories per serving?
2. Which food do you think is the best for your body? Why?

What If

Are most serving sizes we eat the same as shown on the labels? Do we eat bigger or smaller serving sizes?

Why You Need Water

Let's Find Out
- How your body uses water
- How you can replace the water your body uses

Words to Know
sweat
dehydration

The Big QUESTION
Why do you need to drink water?

Using Water

All living things need water. You need water to live. Most of your body weight is water.

Your body uses water in many ways. Water carries nutrients through your body. It helps your body break down food and cools you off when you are hot. Water also takes waste out of your body. You can live longer without food than without water.

You lose water all day long. You lose water when you **sweat.** Sweat helps your body keep cool.

Replacing Water

You need to replace the water that you lose each day. Some water may be replaced with foods. Fruits and vegetables contain lots of water. When you eat an orange, you are replacing water in your body.

The best way to get water into your body is to drink liquids. Water and milk are the best for you. But you can also drink some juices. You should drink six to eight glasses of water each day.

Sometimes you need even more water than usual. When the weather is very hot, or if you are exercising, you sweat more. Your body loses more water. You must replace it. You should drink water before, during, and after exercise, even if you do not feel thirsty.

Not having enough water is called **dehydration**. It can make you sick.

CHECKPOINT

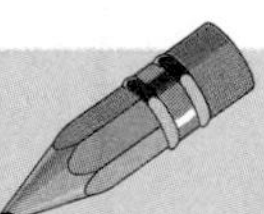

1. How does your body use water?
2. How can you replace the water that your body uses?

Why do you need to drink water?

ACTIVITY

Recording Water Intake

Find Out

Do this activity to see how much water you drink.

Process Skills

Communicating
Measuring
Interpreting Data

WHAT YOU NEED

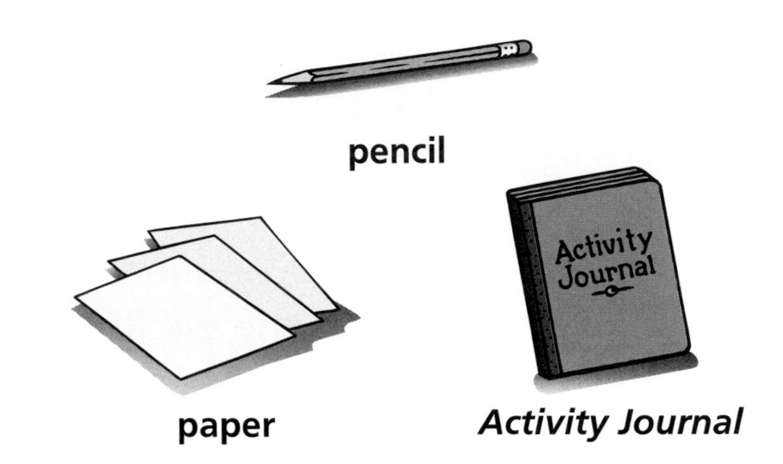

pencil

paper

Activity Journal

WHAT TO DO

1. Keep a water chart for one day. Answer the following questions:
   - When did you drink?
   - Where did you drink?
   - How much did you drink?
   - Were you exercising?
   - How hot was the weather?

2. **Write** down every time you have a drink of water.

3. **Measure** the water you drink by full glass amounts. Sometimes you may drink only half a glass and other times you may drink a glass and a half.

What Happened

1. How many times did you get a drink of water each day?
2. How much did you usually drink each time?

What If

When do you need to drink more water?

Review

What I Know

Choose the best answer for each sentence.

Food Guide Pyramid	**dehydration**
diet	**nutrients**
sweat	**serving**
energy	

1. Your __________ is what you eat regularly.

2. The __________ tells you the kinds and amounts of healthful foods your body needs each day.

3. __________ is your ability to work or play.

4. When you exercise, you lose water through __________.

5. Not having enough water is called __________.

Using What I Know

1. How are these children losing water?

2. How can they quickly replace water?

3. What kind of weather do you think they are having? Why do they need more water in this kind of weather?

4. What activity are the children doing? Why do they need more water during this activity?

5. What might happen to them if they do not replace the water they lose?

For My Portfolio

Write a story about someone who did not eat healthful foods. Tell how you convinced him or her to eat healthful foods and drink plenty of water.

Unit Review

Telling About What I Learned

1. Your senses and sense organs help you describe the world around you. Name one sense organ and tell how it works.
2. Teeth help you chew, talk, and look your best. Name two ways you can take care of your teeth.
3. A healthful diet helps you grow and stay well. Tell about the food groups you should eat from the most and least.

Problem Solving

Use the picture to help answer the questions.

1. What would your eyes and skin tell you about this place? What would they warn you to stay away from?
2. If you came here, what kind of food would you want to bring with you? What else would you need?

Something to Do

Suppose you had friends who didn't take care of their teeth. Using props such as a toothbrush, toothpaste and dental floss, tell them what they should and should not do. Also tell them the reasons why.

Reference

Almanac

Animal Life Cycles

Life Cycle of a Butterfly

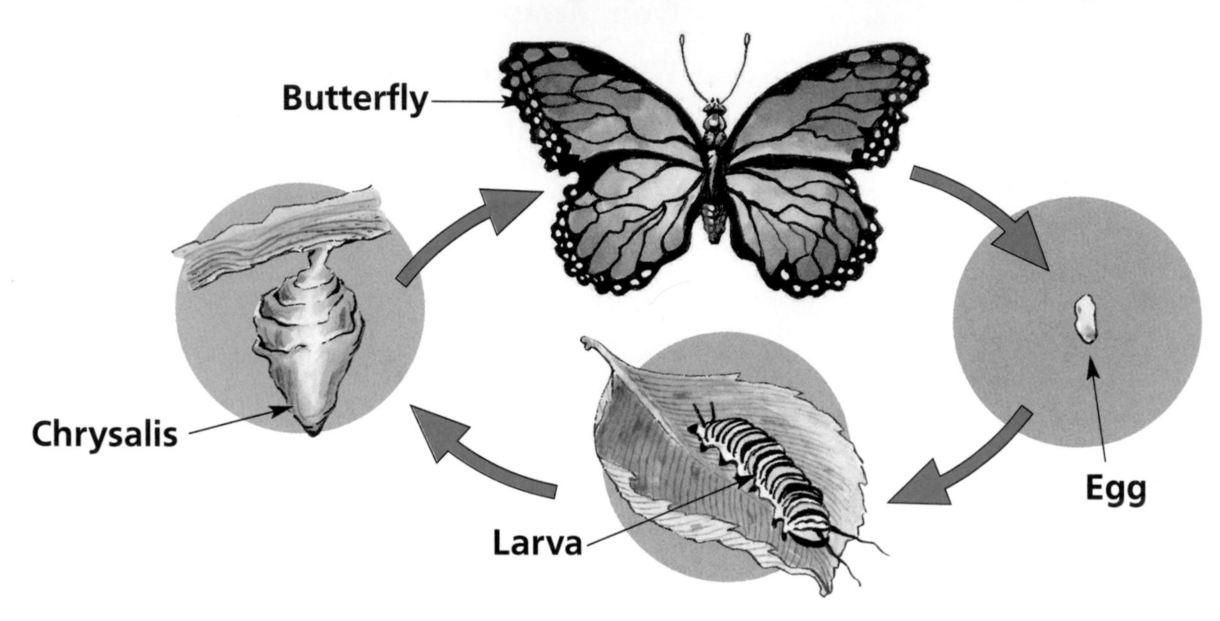

Life Cycle of a Frog

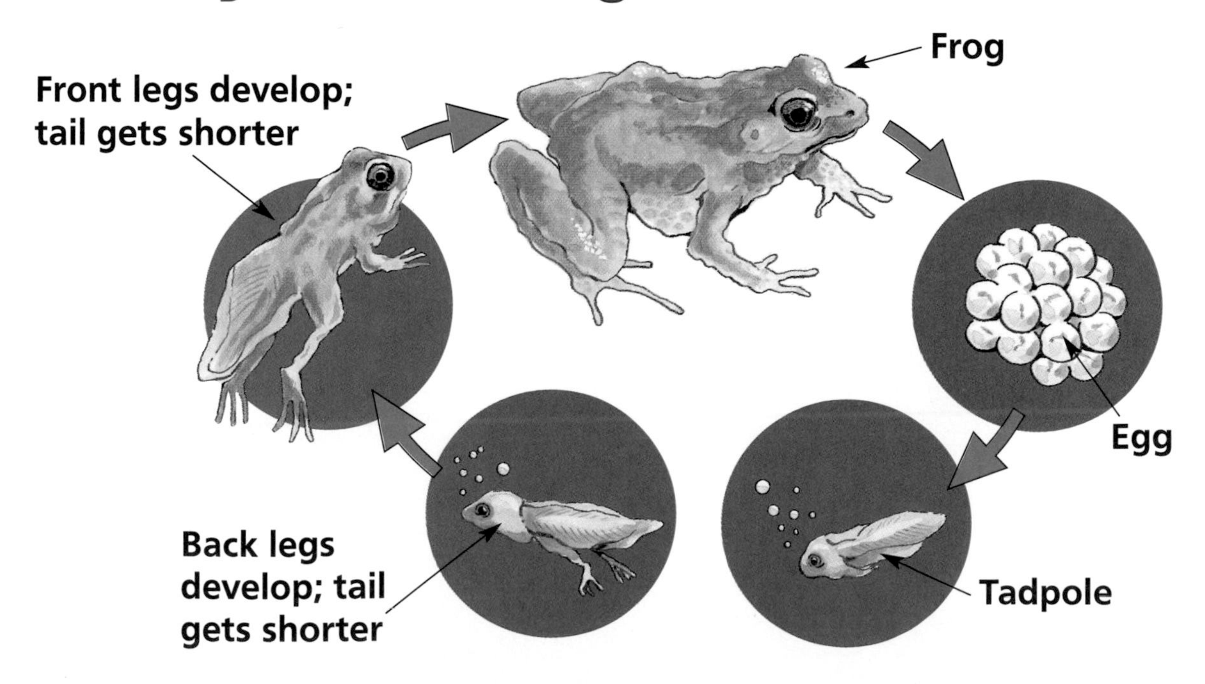

Life Cycle of a Bird

Life Cycle of a Horse

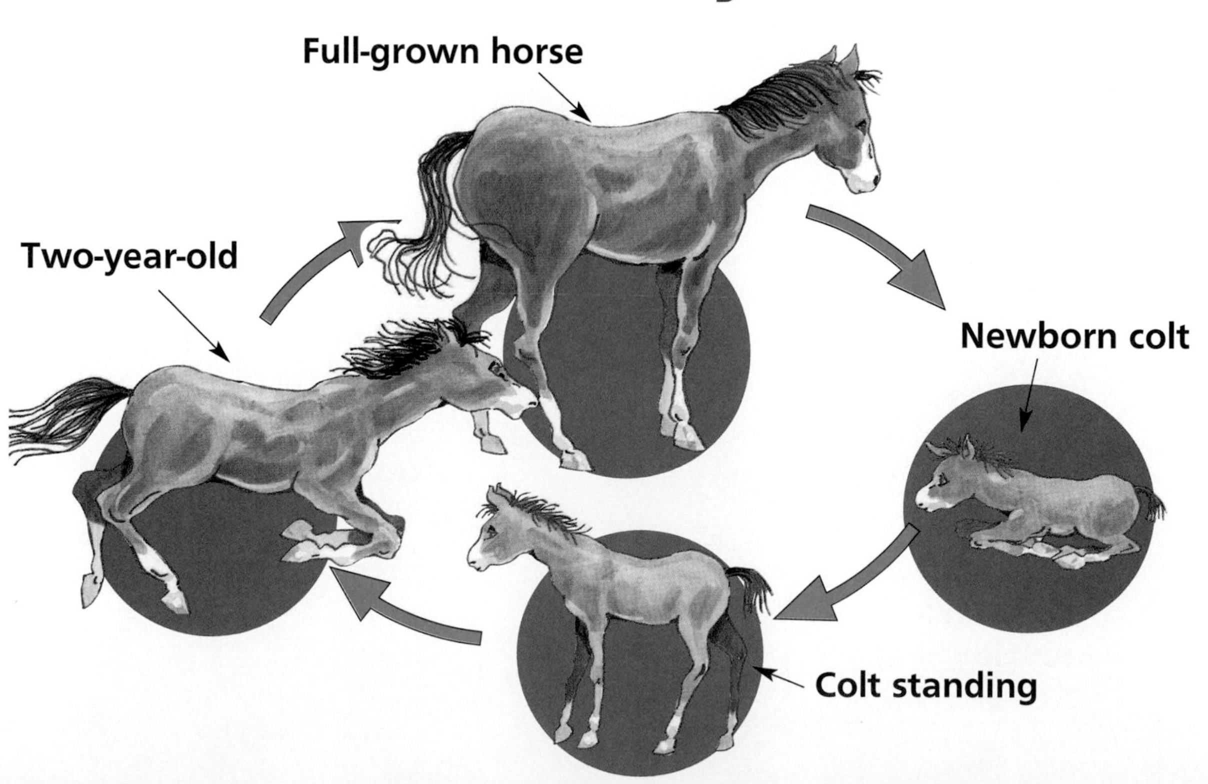

Musical Instruments from Around the World

Congas are drums. The players strike the drums with their hands.

The Veracruz harp is plucked by a person who is standing.

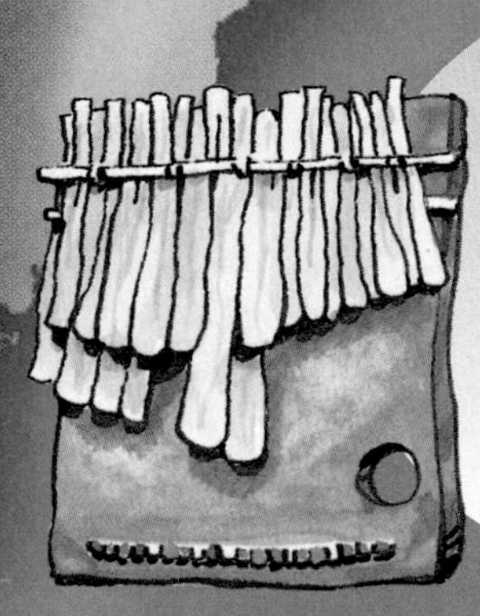

The mbira has metal strips that are attached to a sound board. Players pluck the metal strips.

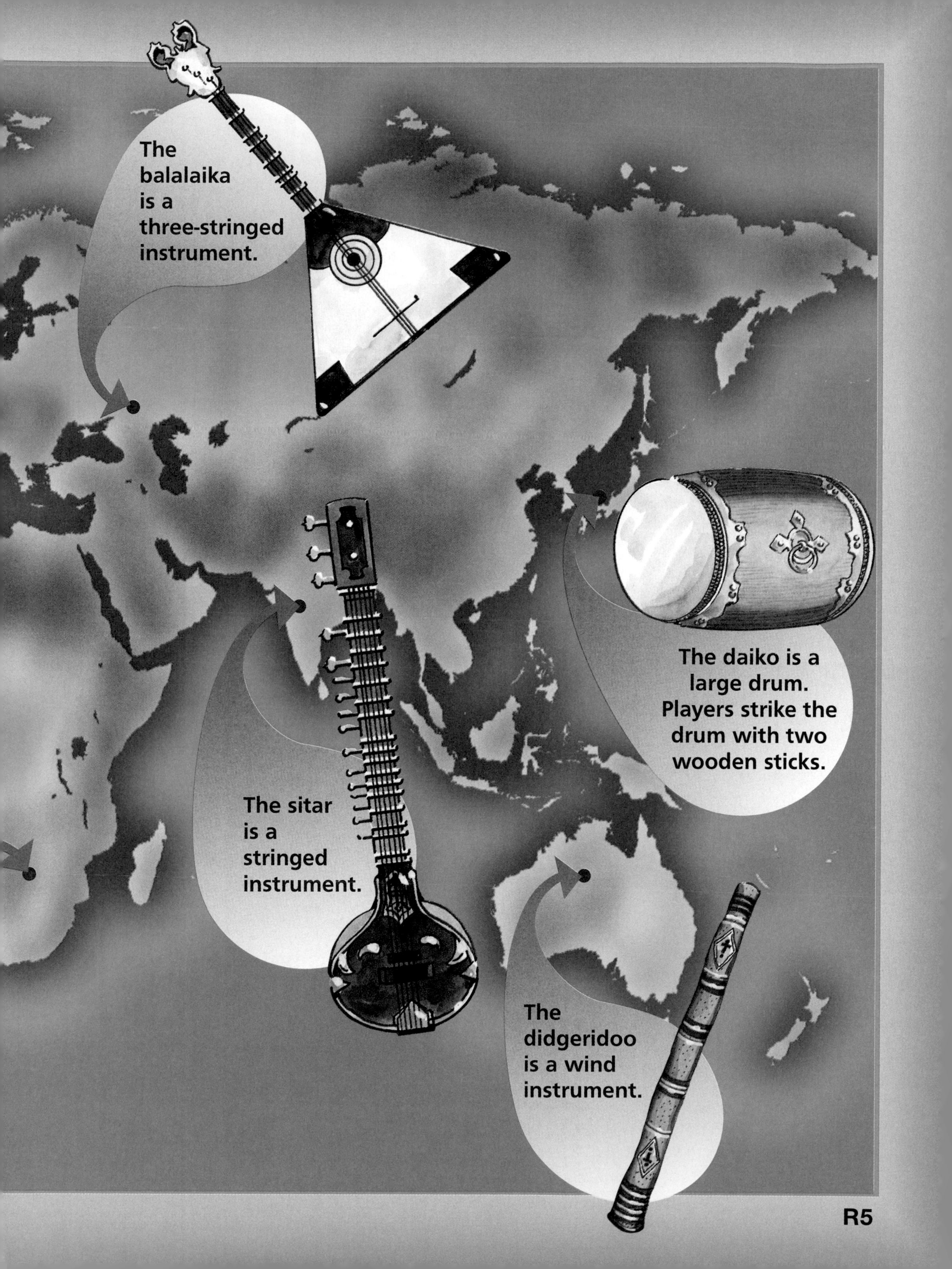
The balalaika is a three-stringed instrument.
The daiko is a large drum. Players strike the drum with two wooden sticks.
The sitar is a stringed instrument.
The didgeridoo is a wind instrument.

What Can You Recycle?

Recycling Facts

- Recycling just one aluminum can saves enough energy to run your TV for three hours.
- If everyone in the United States recycled their Sunday newspaper, it would save 500,000 trees every week.
- Recycling just one glass bottle saves enough energy to burn a lightbulb for four hours.
- Making cans from recycled aluminum uses 90 percent less energy than making cans from scratch.

The Brain

Different parts of the brain do different jobs. Touch, movement, sight, and hearing are all controlled by different parts of the brain. Nerves throughout the body send messages from the eyes, ears, tongue, nose, and skin to the brain. The brain sorts the messages. Then, the brain decides what to do and sends nerve signals back to other parts of the body.

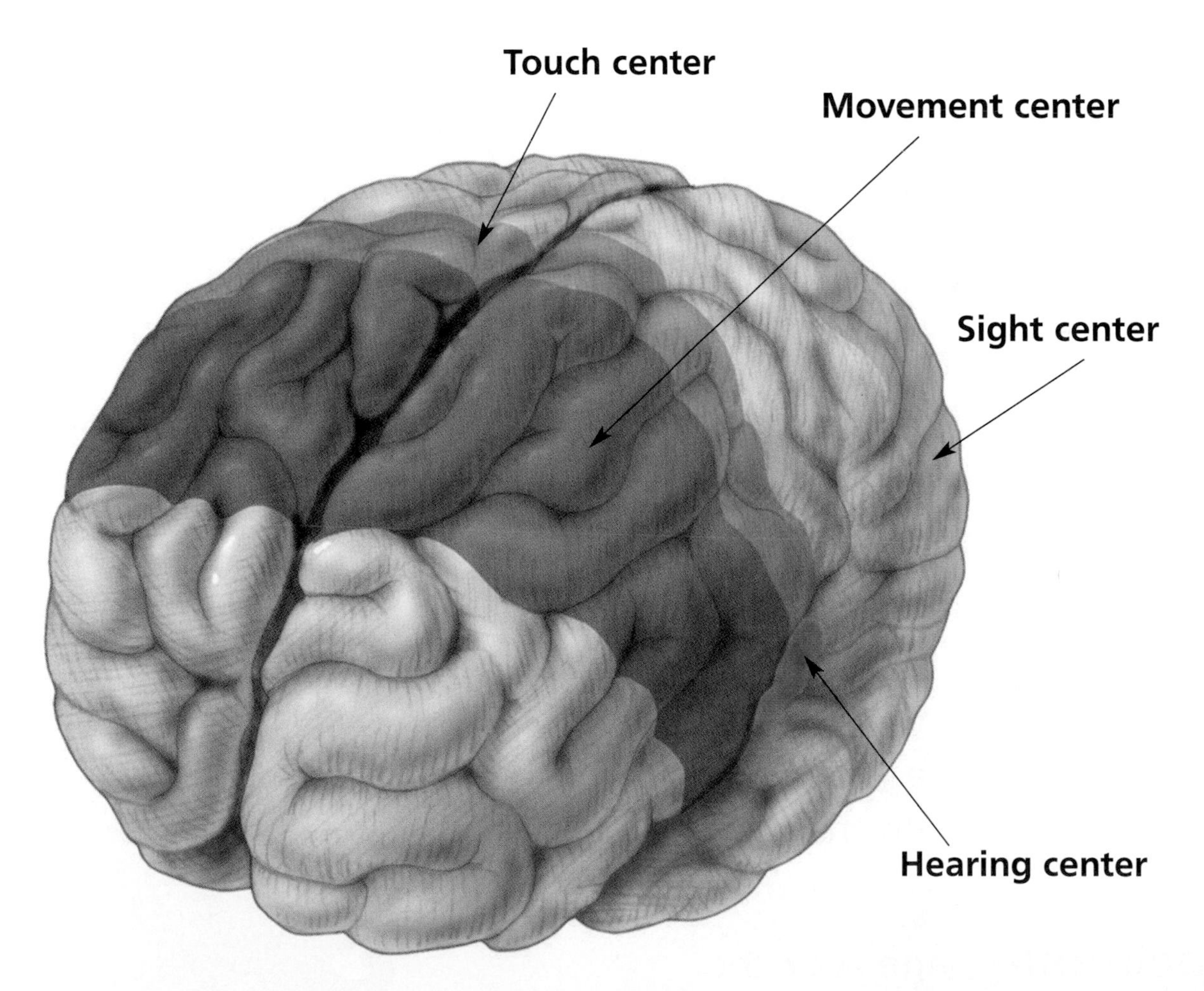

Glossary

A

abdomen page A27 the rear part of an insect's body

absorb page C49 to take in light

antennae page A27 /an ten′ ē/ the pair of long, thin body parts that insects use to learn about the world around them

axis page B50 the imaginary line that runs through the middle of Earth, from the north pole to the south pole

B

baby teeth page D28 the first set of teeth that people have

backbone page A27 the set of bones that run down the center of the back of some animals

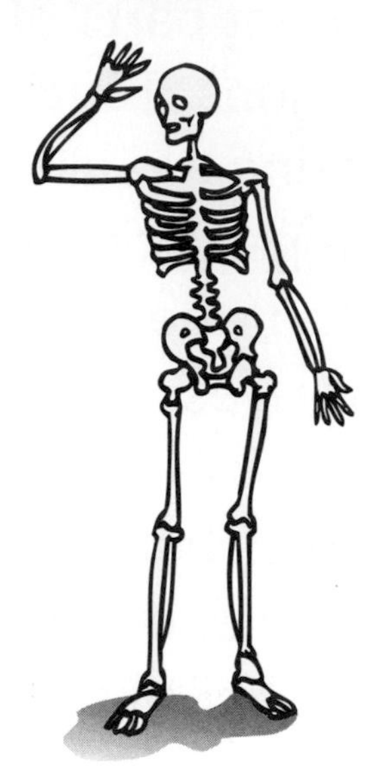

brain page D12 the body part that makes sense of messages sent by the sense organs along nerves

C

cavity page D35 a hole in a tooth caused by tooth decay

climate page B38 what the weather of a place is usually like

Glossary

conductors page C61
materials that heat can move through easily

conservation page B18
using resources wisely

crowding page A18 when plants are too close together and do not get enough water and light

crown page D28 the part of a tooth that can be seen

crust page B26 the top layer of Earth

D

danger page D16
something that may cause harm or injury

defenses page A62 body parts or behaviors that protect living things

dehydration page D65 not having enough water

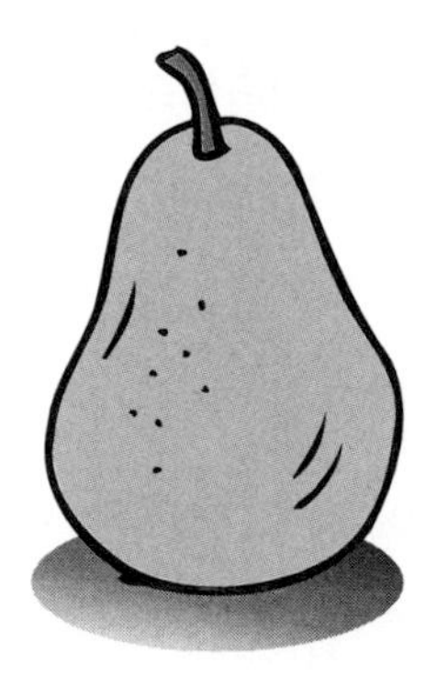

diet page D48 the food that you usually eat

dinosaurs page B38
animals that lived on Earth millions of years before people

direction page C11 the line along which something moves or lies

E

eardrum page D10 the thin piece of skin stretched like the top of a drum inside the ear that vibrates when hit by sound waves

ears page D5 body parts that people use to hear

Glossary

electricity page C54 a form of energy that is used for light and heat

embryo page A10 a new, tiny plant that is part of a seed

energy pages B12, D56 the ability to do work; your ability to work or play

environment page B16 the air, water, soil, and all the other things that surround us

erosion page B35 the movement of rocks and soil by rainwater, streams, and wind

evaporation page C62 water changing from a liquid into water vapor

eyes page D5 body parts that humans use to see

F

filling page D41 a substance used to fill a cavity in a tooth

floss page D33 to clean between the teeth with a string called dental floss

flowers page A6 the parts of some plants that make fruit and seeds

Glossary

fluoride page D39 /flōr′ īd′/ a substance found in most drinking water and toothpastes that helps make teeth strong and less likely to get cavities

Food Guide Pyramid page D50 a guide to healthful eating that includes food groups and suggested servings

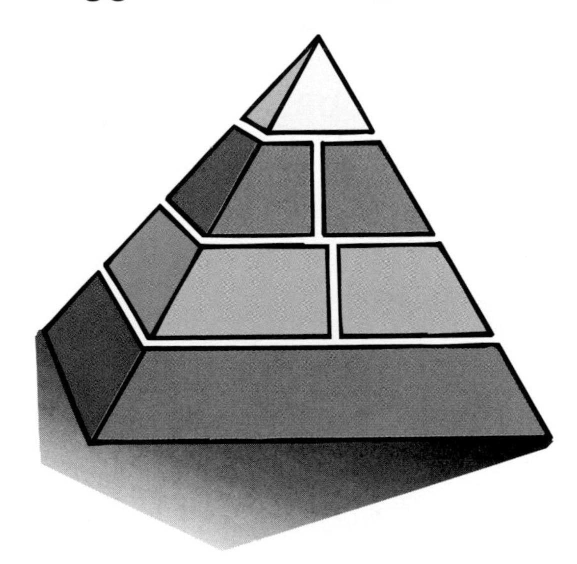

forces page C5 pushes and pulls

fossil fuels page B12 natural resources such as coal, oil, and natural gas that are used for heat, electricity, and transportation

fossils page B40 hardened remains or traces of an animal or plant that once lived

friction pages C12, C57 force caused by two things rubbing together that can slow objects down

fruit page A10 the plant part that contains seeds

fuel page C54 something burned to make heat

full moon page B59 moon phase when you can see all of the lighted side of the moon

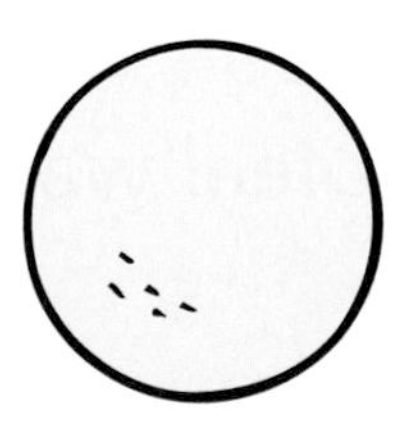

G

germinate page A11 to begin growing from a seed; to sprout

Glossary

globe page B62 a round ball that has a map of Earth on it

gravity page C6 the force that pulls objects toward Earth

gum page D28 the soft, pink flesh in your mouth that surrounds the bottom of the teeth

H

habitats page A56 the places where plants and animals live

heat page C54 what makes an object feel warm

horizon page B48 the line in the distance where the sky meets Earth

I

insects page A27 small animals without backbones that have bodies that are divided into three main parts and have six legs

insulators page C61 materials that help keep heat from passing through

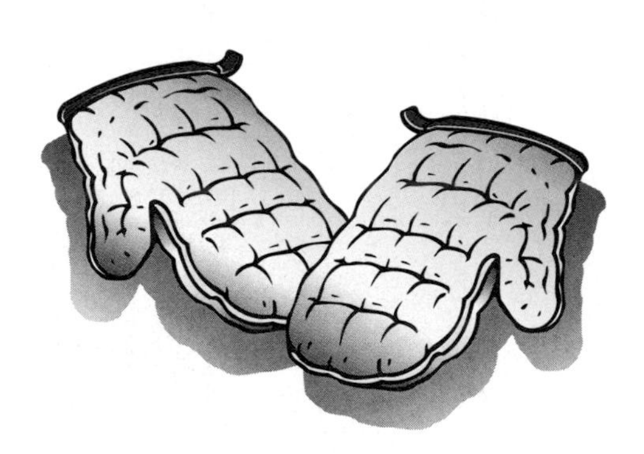

L

larva page A34 the newly hatched form of some insects that spends most of its time eating

leaves page A4 the plant parts in green plants that make food

Glossary

life cycle page A10 how a living thing begins its life, grows, and makes new living things like itself

light page C48 the form of energy that makes it possible for us to see

loud page C32 having a high volume

M

machines page C16 tools used to apply a force

magnets page C18 pieces of stone or metal that attract some metal objects

matter page C62 anything that takes up space and has weight

mimicry page A63 when an animal is protected because it looks like a more dangerous animal

minerals page B28 nonliving materials that come from Earth

motion page C4 when something changes position

music page C40 a pleasing combination of sounds

musical sounds page C40 pleasing sounds

N

natural resources page B4 materials from Earth that plants and animals depend on for life

Glossary

nerves page D12 pathways that carry messages between the brain and other parts of the body

new moon page B59 moon phase when all of the lighted side of the moon faces away from Earth

noise page C35 unpleasant sounds

nose page D5 the body part used for smelling

nutrients pages A5, D52 substances in soil that help plants grow; substances in food that help people work and play

O

orbit page B54 the path that an object in space follows as it moves around another object

organs page D5 body parts that do special jobs

P

permanent teeth page D28 the teeth that grow to replace baby teeth

phase page B58 the lighted part of the moon that can be seen from Earth

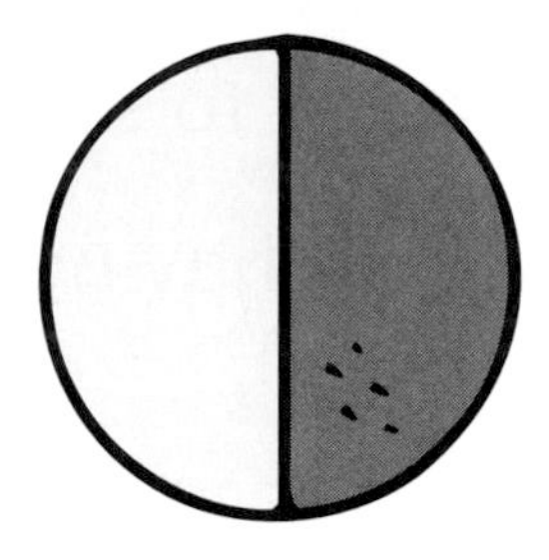

pitch page C38 the highness or lowness of a sound

poison page A62 a substance that can harm a living thing

poles page C18 the ends of a magnet

pollen page A10 powdery material, made by plants with flowers, that makes seeds

pollution page B17 harmful materials that make the air, water, or soil dirty

position page C4 the place where something is

proboscis page A50 /prə bä′ səs/ the long, hollow mouth part that butterflies use to get food from deep inside flowers

protect page D18 to keep from harm

pupa page A34 the form of some insects, after the larva stage and before the adult stage, that does not move

R

recycle page B19 to make a new object out of an object that has been used

reduce page B18 to use less

reflects pages B57, C49 bounces or turns back light

reuse page B19 to use something again

revolves page B54 moves in a path around another object

rock page B26 a substance that is made of minerals and is part of Earth's crust

root page D28 the part of the tooth that is hidden below the gum and holds the tooth in place

roots page A5 the plant parts that hold plants in the soil and take in water and nutrients

rotation page B50 the turning of an object on its axis

S

safe page D18 protected from harm or danger

sand page B33 tiny, loose grains of crushed rocks

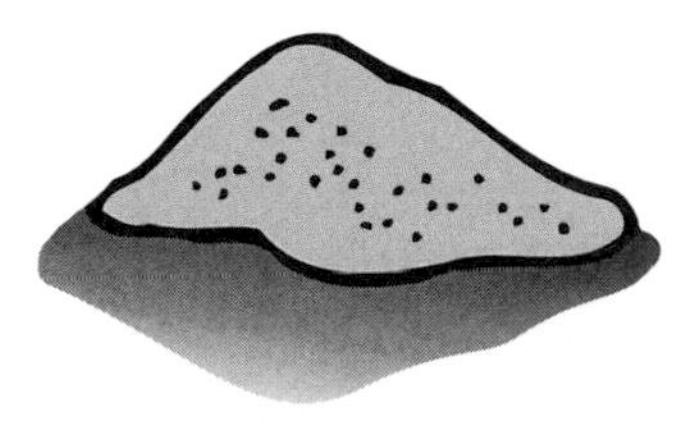

season page B63 a certain time of year with a particular kind of weather

seeds page A6 the parts of some plants from which new plants grow

senses page D4 how people learn about the world; sight, hearing, touch, taste, smell

serving page D52 a suggested amount of food

shadow pages B49, C51 a dark place that is made when light is blocked by an object

skin page D5 the body part that senses touch

soft page C32 having a low volume

Glossary

soil page B34 the part of Earth where plants grow

sound waves page C26 vibrations that can be heard

speed page C10 how fast an object moves

spores page A13 the plant part from which ferns and mosses grow

state page C62 a form of matter; solid, liquid, gas

stem page A5 plant part that carries water to the leaves

sunrise page B48 when the sun seems to move above the horizon

sunset page B48 when the sun seems to move below the horizon

survive page A48 to continue to live

sweat page D63 to make and give off a salty fluid through the skin

Glossary

T

taste buds page D11 the small bumps on the tongue that sense taste

temperature page C65 a measure of how hot or cold something is

thermometers page C65 tools used to measure temperature

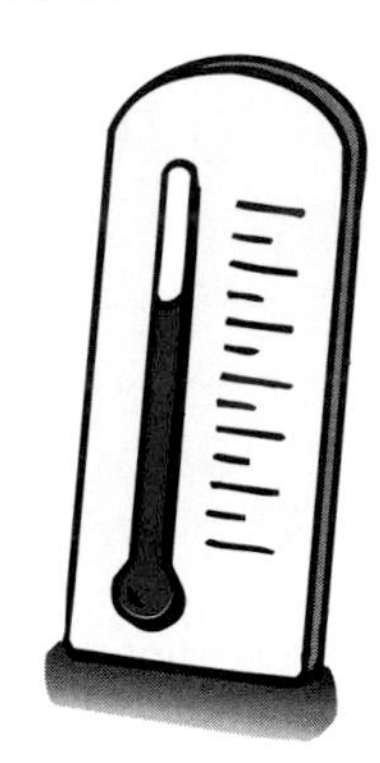

thorax page A27 the part of an insect's body that is between the head and the abdomen

tongue page D5 the body part used for tasting

tooth decay page D34 rotting of the teeth

tooth X ray page D40 a kind of picture that shows the inside of teeth

V

variations pages A16, A38 differences between the same kinds of living things

vibrate pages C26, D11 to move back and forth

vibrations page C26 back-and-forth movements

volume page C33 the loudness or softness of a sound

W

warning page D16 a notice that tells of a danger

water cycle page B6 the movement of water in a cycle from Earth to the clouds and back to Earth

water vapor pages B6, C62 the gas made when liquid water is warmed

weathering page B32 breaking down or wearing away of rock by wind and water

Index

A

Index

B

C

D

Index

Index

Index

Index

W

Credits

Covers, Title Page, Unit Openers, ©VIREO/Academy of Natural Sciences Philadelphia; **iv (t),** ©David Boyle/Earth Scenes, **(b),** ©Don & Pat Valenti/DRK Photo; **vi (t),** ©DiMaggio/Kalish/Peter Arnold, Inc., **(b),** ©Michael P. Gadomski/Photo Researchers, Inc.; **vii,** ©NASA/PhotoTake NYC; **viii (t),** ©Thomas Zimmermann/Tony Stone Images, **(b),** ©Phil Degginger/Tony Stone Images; **ix,** ©Felicia Martinez/PhotoEdit; **v,** ©Kim Heacox Photography/DRK Photo; **x (t),** ©Dan Ham/Tony Stone Images, **(b),** ©Mary Kate Denny/PhotoEdit; **xi,** ©M. Bridwell/PhotoEdit; **xii, xiii,** ©Matt Meadows; **xiv,** ©KS Studios; **xv,** ©Matt Meadows; **A2-A3,** ©David Boyle/Earth Scenes; **A4,** ©Han Reinhard/OKAPIA/Photo Researchers, Inc.; **A5,** ©First Image; **A6 (t),** ©Doug Wechsler/Earth Scenes, **(bl),** ©John Gerlach/Earth Scenes, **(br),** ©Phyllis Greenberg/Earth Scenes; **A7 (t),** ©David Cavagnaro/DRK Photo, **(b),** ©Matt Meadows; **A8, A9,** ©Matt Meadows; **A10, A11,** ©Runk/Schoenberger From Grant Heilman Photography; **A12 (t),** ©Michael Gadomski/Earth Scenes, **(b),** ©Runk/Schoenberger From Grant Heilman Photography; **A13 (t),** ©Tom Bean/DRK Photo, **(bl),** ©M.C. Chamberlain/DRK Photo; **A13 (br),** ©Fred Whitehead/Earth Scenes; **A14, A15,** ©Matt Meadows; **A16,** ©Joanne F. Huemoeller/Earth Scenes; **A17,** ©David Cavagnaro/DRK Photo; **A18 (t),** ©Joe McDonald/Earth Scenes, **(bl),** ©Lynn M. Stone/DRK Photo, **(br),** Holt Studios International/Nigel Cattlin/Photo Researchers, Inc.; **A19,** ©Porterfield/Chickering/Photo Researchers, Inc.; **A20, A21,** ©George Anderson; **A23,** ©David Cavagnaro/DRK Photo; **A24-A25,** ©Don & Pat Valenti/DRK Photo; **A26,** ©E.R. Degginger/Animals Animals; **A28 (tl),** ©Paul Berquist/Animals Animals, **(tr),** ©Stanley Breeden/DRK Photo, **(c),** ©Leonard Lee Rue III/DRK Photo, **(b),** ©Zig Leszczynski/Animals Animals; **A29 (tl),** ©Darrell Gulin/DRK Photo, **(tr),** ©T.A. Wiewandt/DRK Photo, **(br),** ©Juan Manuel Renjifo/Animals Animals; **A30, A31,** ©Matt Meadows; **A32,** ©Christopher Bissell/Tony Stone Images; **A33 (t),** ©Arthur C. Smith III From Grant Heilman Photography, **(b),** ©Zig Leszczynski/Animals Animals; **A34,** ©Runk/Schoenberger From Grant Heilman Photography; **A35 (tl, tr),** ©E.R. Degginger/Animals Animals, **(b),** ©Scott Smith/Animals Animals; **A36, A37,** ©Matt Meadows; **A38,** ©Walter Hodges/Tony Stone Images; **A39,** ©Alan & Sandy Carey/Photo Researchers, Inc.; **A40 (t),** ©C.C. Lockwood/DRK Photo, **(c),** ©Runk/Schoenberger From Grant Heilman Photography, **(b),** ©Fred Habegger From Grant Heilman Photography; **A41 (tl),** ©Leen Van Der Slik/Animals Animals, **(tr),** ©A. Shay/Animals Animals; **A42, A43,** ©Matt Meadows; **A45 (l),** ©Peter Baumann/Animals Animals, **(r),** ©Scott Camazine/Photo Researchers, Inc.; **A46-A47,** ©Kim Heacox Photography/DRK Photo; **A48 (l),** ©Doug Perrine/DRK Photo, **(r),** ©M. P. Kahl/DRK Photo; **A49 (tr),** ©Robert Lubeck/Earth Scenes, **(l),** ©Lynwood Chase/Photo Researchers, Inc., **(br),** ©Stephen J. Krasemann/DRK Photo; **A50 (l),** ©Michael Fogden/Animals Animals, **(r),** ©Ben Mikaelsen/Peter Arnold, Inc.; **A51 (t),** ©Zig Leszczynski/Animals Animals, **(b),** ©Patti Murray/Animals Animals; **A52 (t),** ©Max Gibbs/Oxford Scientific Films/Animals Animals, **(b),** ©Jerry Cooke/Animals Animals; **A53,** ©Norbert Wu/DRK Photo; **A54, A55,** ©Matt Meadows; **A56,** ©Pat O'Hara/DRK Photo; **A57 (t),** ©Stephen G. Maka/DRK Photo, **(b),** ©Harry Rogers/Photo Researchers, Inc.; **A58 (tl),** ©Fred McConnaughey/Photo Researchers, Inc., **(r),** ©Zig Leszczynski/Animals Animals, **(bl),** ©Dan Suzio/Animals Animals; **A59 (t),** ©Stephen J. Krasemann/DRK Photo, **(b),** ©Kjell B. Sandved/Photo Researchers, Inc.; **A60,** ©KS Studios; **A61 (l),** ©Matt Meadows, **(r),** ©KS Studios; **A62 (l),** ©Alan & Linda Detrick/Photo Researchers, Inc., **(r),** ©Jim Strawser From Grant Heilman Photography; **A63 (tr),** ©Joe McDonald/DRK Photo, **(cr),** ©Michael Fogden/DRK Photo, **(bl),** ©Don Enger/Animals Animals, **(br),** ©John Gerlach/DRK Photo; **A64 (tl),** ©Stanley Breeden/DRK Photo, **(bl),** ©C. C. Lockwood/DRK Photo, **(br),** ©John Gerlach/DRK Photo; **A65 (t),** ©Dagmar/Earth Scenes, **(b),** ©D. Allen/Animals Animals; **A66, A67,** ©KS Studios; **A69,** ©Stephen J. Krasemann/DRK Photo; **A70,** ©Peter Pickford/DRK Photo; **A71,** ©KS Studios; **B2-B3,** ©DiMaggio/Kalish/Peter Arnold, Inc.; **B4,** ©Tom Prettyman/PhotoEdit; **B5,** ©NASA/TSADO/Tom Stack & Associates; **B8, B9,** ©Matt Meadows; **B10,** ©Myrleen Ferguson/PhotoEdit; **B11 (l),** ©Myrleen Ferguson/PhotoEdit, **(r),** ©A. Ramey/PhotoEdit; **B12,** ©Robert Holmgren/Peter Arnold, Inc.; **B13 (tl),** ©Jeff Lepore/Photo Researchers, Inc., **(tr),** ©Thomas Kitchin/Tom Stack & Associates, **(br),** ©Blair Seitz/Photo Researchers, Inc.; **B14, B15,** ©Matt Meadows; **B16,** ©Matt Meadows/Peter Arnold, Inc.; **B17,** ©Adam Jones/Photo Researchers, Inc.; **B18 (t),** ©KS Studios, **(cl),** ©Leonard Lessin/Peter Arnold, Inc., **(bl),** ©Aaron Haupt; **B19,** ©Tony Freeman/PhotoEdit; **B20, B21,** ©Matt Meadows; **B23,** ©Kenneth H. Thomas/Photo Researchers, Inc.; **B24-B25,** ©Michael P. Gadomski/Photo Researchers, Inc.; **B26,** ©Ken M. Johns/Photo Researchers, Inc.; **B27,** ©N.M. Hauprich/Photo Researchers, Inc.; **B28-B31** ©Matt Meadows; **B32,** ©Peter Arnold, Inc.; **B33 (t),** ©David Smiley/Peter Arnold, Inc., **(b),** ©Spencer Swanger/Tom Stack & Associates; **B34 (t),** ©KS Studios, **(b),** ©Paul Barton/The Stock Market; **B35,** ©Robert Carlyle Day/Photo Researchers, Inc.; **B36, B37,** ©KS Studios; **B40,** ©Alfred Pasieka/Science Photo Library/Photo Researchers, Inc.; **B42, B43,** ©Matt Meadows; **B45,** ©Bob Pool/Tom Stack & Associates; **B46-B47,** ©NASA/PhotoTake NYC; **B48,** ©Dennis Flaherty Photography/Photo Researchers, Inc.; **B49 (t),** ©V.A.F. Neal/Photo Researchers, Inc., **(bl, br),** ©David Youngblood/PhotoEdit; **B51 (l),** ©Richard Hutchings/PhotoEdit, **(r),** ©Will & Deni McIntyre/Photo Researchers, Inc.; **B52, B53,** ©KS Studios; **B54,** ©NASA/Photo Researchers, Inc.; **B57,** ©David Nunuk/Science Photo Library/Photo Researchers, Inc.; **B60, B61,** ©Matt Meadows; **B62,** ©KS Studios; **B64 (t),** ©Kevin Schafer/Peter Arnold, Inc., **(b),** ©Gordon Wiltsie/Peter Arnold, Inc.; **B65 (l),** ©Robert Brenner/PhotoEdit, **(r),** ©Norbert Schafer/The Stock Market; **B66, B67,** ©Ken Karp; **B69,** ©Kevin Magee/Tom Stack & Associates; **B70,** ©Hans Reinhard/OKAPIA/Photo Researchers, Inc.; **B71,** ©Matt Meadows; **C2-C3,** ©Thomas Zimmermann/Tony Stone Images; **C4,** ©Studiohio; **C5 (tr),** ©Tom & DeeAnn McCarthy/The Stock Market, **(bl),** ©Tony Freeman/PhotoEdit, **(br),** ©Nancy Sheehan/PhotoEdit; **C6,** ©Bill Aaron/PhotoEdit; **C7,** ©Larry Ulrich; **C8,** ©Matt Meadows; **C9,** ©Matt Meadows; **C10,** ©Bob Daemmrich; **C11,** ©Ron Kimball; **C12,** ©Michael Newman/PhotoEdit; **C13,** ©B. Seitz/ Weststock; **C14, C15,** ©Matt Meadows; **C16,** ©Doug Martin/Photo Researchers, Inc.; **C17 (tr),** ©Felicia Martinez/PhotoEdit, **(bl),** ©Steve Taylor/Tony Stone Images, **(br),** ©Tony Freeman/PhotoEdit; **C18,** ©Tony Freeman/PhotoEdit; **C19,** ©Terry Vine/Tony Stone Images; **C20, C21,** ©KS Studios; **C23,** ©Matt Meadows; **C24-C25,** ©Phil Degginger/Tony Stone Images; **C26 (bl),** ©KS Studios, **(br),** ©Tim Flach/Tony Stone Images; **C28 (tl),** ©Gary Conner/PhotoEdit, **(tr),** ©David Young-Wolff/PhotoEdit, **(b),** ©Dick Rowan/Photo Researchers, Inc.; **C29 (tl),** ©Christoph Burki/Tony Stone Images, **(cr),** ©D. Wilder/Tom Stack & Associates; **C30,** ©KS Studios; **C32,** ©Michael Newman/PhotoEdit; **C33** ©KS Studios; **C34,** ©Bob Daemmrich Photo, Inc.; **C35,** ©Michael Newman/PhotoEdit; **C36, C37,** ©Matt Meadows; **C38,** ©Jeanne White/Photo Researchers, Inc.; **C39 (tr),** ©KS Studios, **(b),** ©Richard Gross/The Stock Market; **C40,** ©Andrea Brizzi/The Stock Market; **C41,** ©Robert E. Daemmrich/Tony Stone Images; **C42, C43,** ©Matt Meadows; **C45,** ©Cosmo Condina/Tony Stone Images; **C46-C47,** ©Felicia Martinez/PhotoEdit; **C48,** ©Joe Monroe/Photo Researchers, Inc.; **C49 (tr),** ©David Madison, **(br),** ©Matt Meadows; **C50,** ©KS Studios; **C51,** ©Jeffry W. Myers/The Stock Market; **C52, C53,** ©KS Studios; **C54 (tl),** ©Doug Martin, **(tr),** ©Chas Krider, **(cr),** ©Brent Turner/BLT Productions, **(bl),** ©Comstock, **(br),** ©Bill Hickey/The Image Bank; **C55,** ©Merritt Vincent/PhotoEdit; **C56,** ©Mark Burnett/Photo Researchers, Inc.; **C57,** ©Mary M. Steinbacher/PhotoEdit; **C58,** ©Michael Newman/PhotoEdit; **C59,** ©Michael Newman/PhotoEdit; **C60,** ©Pat Lacroix/The Image Bank; **C61 (tr),** ©Felicia Martinez/PhotoEdit, **(br),** ©Jeff Greenberg/PhotoEdit; **C62 (tl),** ©Terry Donnelly/Tom Stack & Associates, **(cl),** ©Michael A. Keller/The Stock Market, **(bl),** ©David Young-Wolff/PhotoEdit; **C63 (t),** ©Michael L. Keller/The Stock Market, **(br),** ©Thomas Brase/Tony Stone Images; **C64,** ©Dan Ham/Tony Stone Images; **C65,** ©David Young-Wolff/PhotoEdit; **C66, C67,** ©Matt Meadows; **C69,** ©Aaron Haupt; **C70,** ©Michael Newman/PhotoEdit; **C71,** ©Matt Meadows; **D2-D3,** ©Dan Ham/Tony Stone Images; **D4,** ©Barbara Stitzer/PhotoEdit; **D5,** ©Paul Barton/The Stock Market; **D6 (tr),** ©Rich Iwasaki/Tony Stone Images, **(bl),** ©Spencer Grant/PhotoEdit; **D7,** ©John Gerlach/Tom Stack & Associates; **D8, D9,** ©Matt Meadows; **D11,** ©Tony Freeman/PhotoEdit; **D13,** ©KS Studios; **D14,** ©Matt Meadows; **D16, D17,** ©David Young-Wolff/PhotoEdit; **D18 (tr),** ©Bonnie Kamin/PhotoEdit, **(bl),** ©Franklin Avery/Tony Stone Images; **D19,** ©Michael Newman/PhotoEdit; **D20, D21,** ©Matt Meadows; **D23,** ©Richard Brown/Tony Stone Images; **D24-D25,** ©Mary Kate Denny/PhotoEdit; **D26,** ©Richard Hutchings/PhotoEdit; **D27 (tr),** ©Nancy Ney/The Stock Market, **(bl),** ©Juan Silva/The Image Bank, **(br),** ©Nancy Ney/The Stock Market; **D28,** ©Myrleen Ferguson/PhotoEdit; **D30, D31,** ©Matt Meadows; **D32,** ©David Young-Wolff/PhotoEdit; **D33,** ©Spencer Grant/PhotoEdit; **D34,** ©SuperStock; **D36, D37,** ©Matt Meadows; **D38,** ©Bill Aron/PhotoEdit; **D39,** ©Michael Newman/PhotoEdit; **D40 (t),** ©Kourosh Behbahani, **(bl),** ©Michael Newman/PhotoEdit; **D41,** ©Rick Gayle/The Stock Market; **D42, D43,** ©Matt Meadows; **D45,** ©David Young-Wolff/PhotoEdit; **D46-D47,** ©M. Bridwell/PhotoEdit; **D48,** ©Spencer Grant/PhotoEdit; **D49 (tr),** ©Tony Freeman/PhotoEdit, **(cl),** ©David K. Crow/PhotoEdit, **(cr, bl, br),** ©Felicia Martinez/PhotoEdit; **D51,** ©Michael Newman/PhotoEdit; **D52, D53,** ©David Young-Wolff/PhotoEdit; **D54,** ©Matt Meadows; **D55,** ©Kelvin Murray/Tony Stone Images; **D56,** ©Lori Adamski Peek/Tony Stone Images; **D58,** ©David Young-Wolff/PhotoEdit; **D59,** ©Mary Kate Denny/PhotoEdit; **D60,** ©Matt Meadows; **D61,** ©Michael Newman/PhotoEdit; **D62,** ©Fotomorgano/The Stock Market; **D63 (tr),** ©David Young-Wolff/PhotoEdit, **(bl),** ©Michael Newman/PhotoEdit; **D64 (bl),** ©Myrleen Ferguson/PhotoEdit, **(br),** ©Spencer Grant/PhotoEdit; **D65,** ©Tom Prettyman/PhotoEdit; **D66,** ©Matt Meadows; **D67 (tr),** ©Matt Meadows, **(cr),** ©Mary Kate Denny/PhotoEdit; **D69,** ©94DiMaggio/Kalish/The Stock Market; **D70,** ©Jeff Greenberg/PhotoEdit; **D71,** ©Matt Meadows; **R6 (tl),** ©Michael P. Gadomski/Bruce Coleman, Inc., **(tr),** ©C. Bradley Simmons/Bruce Coleman, Inc., **(cl),** ©SuperStock, **(cr),** ©G&G Schaub/SuperStock; **(cbl),** ©Peter French/Bruce Coleman, Inc., **(cbr),** ©Ken Sherman/Bruce Coleman, Inc., **(bl),** ©SuperStock, **(bcl),** ©Phil Degginger/Bruce Coleman, Inc.